THE CARBON RUSH

THE CARBON RUSH

Amy Miller

Red Deer Press

5 4 3 2 1

Published in Canada by Red Deer Press
195 Allstate Parkway, Markham, ON, L3R 4T8
www.reddeerpress.com

Published in the United States by Red Deer Press
311 Washington Street, Brighton, Massachusetts, 02135

Cover layout and text design by Daniel Choi
Cover art courtesy of KINOSMTH INC
All photos courtesy of Carbon Rush Inc., from the feature documentary THE CARBON RUSH, director of photography PAUL KELL.

We acknowledge with thanks the Canada Council for the Arts, and the Ontario Arts Council for their support of our publishing program. We acknowledge the financial support of the Government of Canada through the Canada Book Fund (CBF) for our publishing activities.

Canada Council for the Arts

Conseil des Arts du Canada

Library and Archives Canada Cataloguing in Publication
Miller, Amy
The Carbon Rush / Amy Miller.
ISBN 978-0-88995-479-3
1. Carbon offsetting. 2. Emissions trading. I. Title.
HC79.P55M55 2013 363.738'746 C2013-900606-0

Publisher Cataloging-in-Publication Data (U.S.)
Miller, Amy.
The Carbon Rush / Amy Miller.
[180] p. : col. ill. ; cm.
Summary: Based on an award-winning documentary, this book focuses on the real effects of on carbon trading and how it is ravaging the world's poor and their environments.
ISBN: 978-0-88995-479-3 (pbk.)
1. Carbon offsetting. 2. Climatic changes -- Economic aspects. 3. Pollution -- Economic aspects. 4. Capitalism. 5. Globalization.
I. Title.
363.738/746 dc23 HC79.P55M455 2013

Printed and bound in China by Sheck Wah Tong Printing Press Ltd.

This book is dedicated to all of those who fight tirelessly for social and environmental justice around the world. We would like to give our heartfelt special thanks to those communities who shared their stories in the making of THE CARBON RUSH.

Sincerely,
Amy Miller and Byron A. Martin
producers of *The Carbon Rush*

Contents

November 2013 saw the Warsaw Climate Change Conference, the COP19 summit organized by the United Nations, come and go without the global-environmental-governmental elites doing anything worth mentioning to solve any of the pressing socio-environmental problems. Sadly, this is not a surprise, given their track record since the first Rio conference on sustainability in 1992. The previous gathering to fix the climate crisis, the COP 18, was in Qatar, a country with massive oil and natural gas reserves. This fact should lead us to think not just about oil, but also about control and power, imperial might, and the perpetuation of that might at any cost and how the global oil economy is destroying our basic needs for survival.

Right now the climate crisis and peak oil are converging. As it is, the predominant belief that has come out of North America about this very real climate crisis does *not* include focusing resources so we can move out and away from a fossil fuel economy. Instead, it states that the climate crisis is a problem that can be solved through minor changes in consumption habits and massive international trading ventures, which do little but reinforce the existing system. There should be no surprise then that to date, there have been over three hundred billion dollars worth of carbon transactions worldwide—but without any of the desired reduction in greenhouse gas emissions.

Climate change, or as I prefer, climate chaos, is confirmation of the limits faced by a system of infinite growth on a finite planet. The methods proposed and being put in place by the international community—the United Nations—to rectify the problem are market-based mechanisms, primarily cap-and-trade

INTRODUCTION

carbon markets and offsetting schemes. Signatory states to the UN Kyoto Accord are implementing these measures to various degrees and with varying speed. These methods are inherently flawed as they function within an economic framework that feeds only blind profit, rather than recognize the framework itself as a root problem. By focusing on carbon instead of the flows of capital responsible for its emission, policy makers are misleading the effects, with the system that produces them. This failure reveals the contradictions built in a system that is responsible for causing this crisis.

The Carbon Rush came out of many diverse discussions that occurred during the nine-week, forty-four screening tour I did with my previous documentary, *Myths for Profit: Canada's Role in Industries of War and Peace.* After the screenings, all kinds of different people came up to me to talk about what currently fuels wars around the world. Questions of resource control, primarily over fossil fuel and

water, were raised at each screening and this inevitably turned into a discussion on the impending shortage of both of these incredible resources and how our unending appetite for oil is fueling climate chaos. At the time, I had been living with the talented director Shannon Walsh who had just finished *H2Oil*, a critical documentary looking at the realities of the tar sands, so these ideas had already been swarming around in my head. Carbon markets and offset projects? How come nobody knew anything about this? It was then that I realized the need for something like *The Carbon Rush.*

From various discussions it became evident that people were very unfamiliar with the concept of offsets—that is, the purchasing of carbon credits that support a project somewhere else which in turn allows the carbon-credit purchaser at home to continue polluting. People *here* had no idea how these offset projects were impacting people *there*, in places where these projects were set up. And to be honest, neither did I. But I had a sneaking suspicion that it wasn't very good.

When I returned to Montreal in May 2009 I was profoundly impacted by these discussions and the connections drawn around what I saw as a new form of expansionism/colonialism hidden behind the cunning agenda of "solving the climate crisis." I quickly began researching the subject and talking to people who looked at topics that touched on carbon markets and offsets. This work confirmed my suspicions—there was currently no documentary on this crucial subject! While the issue is vitally important it is surprisingly only in the last year or two that it started to garner some of the attention it requires.

Climate chaos stands alongside the current political-economic crisis and the present-day energy, food and water

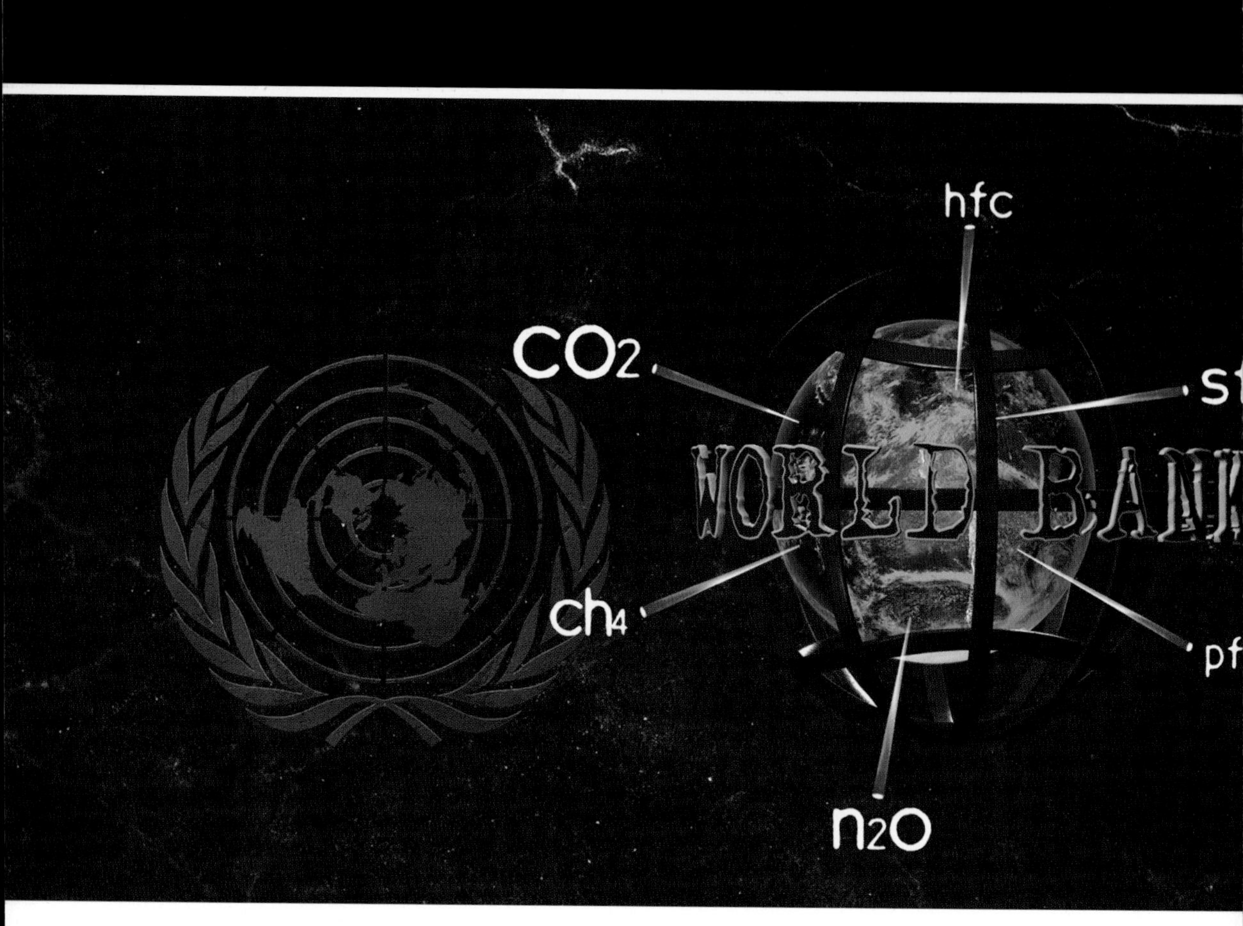

crises as being caused and worsened by an economy that feeds off of endless growth. The ruling elites—those who run the factories, who are in the backrooms of power, those who benefit most from the profits—these people are consequently seeking to legitimize a system which is the root cause of all of these socio-ecological crises, and they are using "crisis management" as a smokescreen opportunity to reassert the system that caused the problem in the first place.

They are using the same approach, policy strategy, and environmental discourse of "ecological modernization" to push false solutions such as carbon trading, carbon capture and storage, agrofuels, and nuclear power, all of which continue the concentration of political and economic power that first got us into this mess. Despite ever increasing destruction of our planet's ecosystems and widening inequality between rich and poor, the global ruling elites consistently chose self-serving short-term gains that support economic growth. In other words, it's business as usual.

At the 2010 NATO summit in Lisbon, military representatives from around the world discussed their solutions to the impending disasters born from climate chaos: militarized borders and stricter internal security measures like biometric IDs and surveillance. Already 300,000 people die every year, nearly all of them in the Global South, due to the results of climate change: desertification, droughts, more violent storms, the increasing spread of tropical diseases, and crop failure. Human populations are already beginning to migrate on an enormous scale in search of means to survive. And as a sharp decrease in agricultural productivity caused by changes in temperatures

coincides with a peaking human population, things will get beyond critical.

Currently, our political institutions are unable to respond to the scale of this challenge due to their commitment to the existing political/corporate/economic agenda. Therefore, solutions *must* come from beyond traditional political institutions and from the people themselves. Through an emancipatory transformation of social relations, often referred to as "climate justice," people worldwide are becoming more informed. They are asking questions, and demanding answers. Social groups are collecting and gathering steam and bringing new solutions to our current climate crisis. Thus, the hope for the survival of future generations relies on us, and all those movements worldwide that push projects and solutions for global climate and economic justice.

I believe that knowledge leads to transformative change. When people know what is happening they are concerned and want to take action, not only to create a more just and equitable world, but also to have a planet humans can live and survive on. It is basic self-interest. We are not talking about an issue that people can simply dismiss, because these problems impact on each and every one of us.

***

Much of the way I present ideas in my documentaries is rooted in Paulo Freire's beliefs surrounding popular education. Freire emphasized that the goal for education is not to create what he called "neutral" functional conformity, but, instead, to help create a critical questioning awareness. This is the goal of my documentaries and by extension the goal of this book. My documentaries are inspired by the critical

WORL

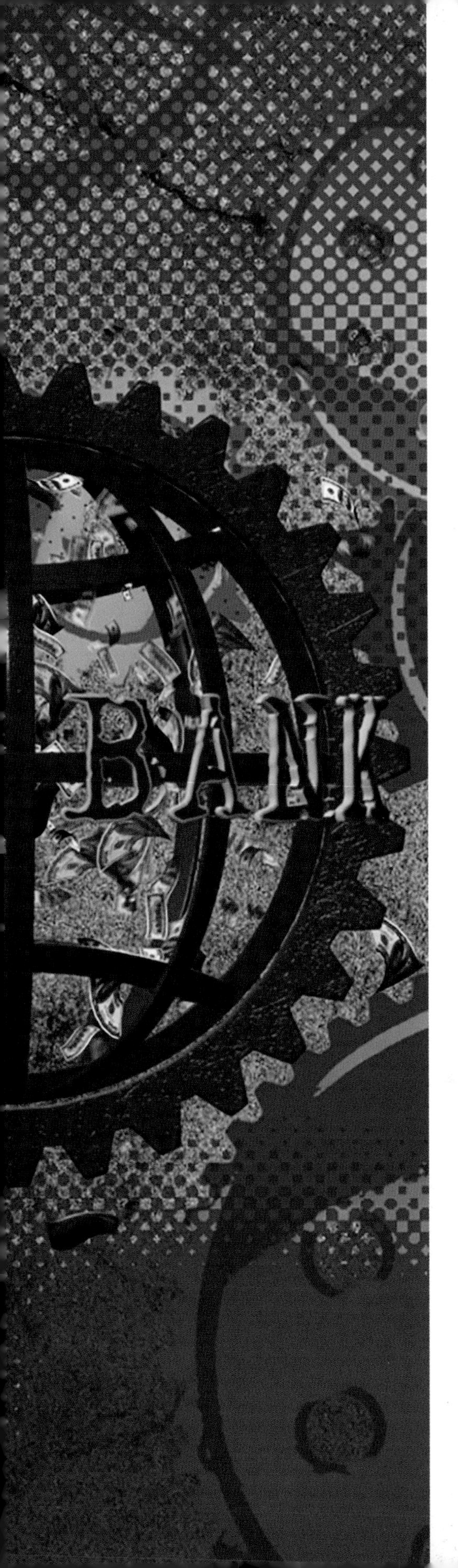

pedagogy movement that Freire and others such as Henry Giroux have developed: guided by passion and principle, to help audiences develop consciousness of freedom, recognize authoritarian tendencies, connect knowledge to power, and develop the ability to take constructive action.

It is my hope that *The Carbon Rush* can transcend national identities and show that solving the climate crisis cannot be simply a transfer of responsibility from the Global North onto the Global South. I also hope that audiences and readers will hear —maybe for the first time—the voices of those that have been most silenced, and victimized by this climate chaos. Only by listening can we ensure that climate justice is a collective endeavor.

We can see the flowering of a global movement which articulates climate justice along with a multitude of related struggles. From Copenhagen to Cochabamba and beyond, the climate justice movement is growing, fighting back, and refusing to back down. I hope that *The Carbon Rush* is one tool that people can use to further the movement.

Amy Miller

CARBON MARKET

How It Works

By Amy Miller
Social Activist, Writer, Film Maker

To date, there has been over three hundred billion dollars of transactions worldwide from the carbon credit and offset market.

There are currently over 5,000 projects registered in the United Nations carbon market. What do they all have in common?

They are all receiving carbon credits for offsetting pollution created somewhere else.

Carbon credits allow polluters in the affluent Global North to continue their activities unchecked as long as someone, somewhere in a developing nation, is simultaneously doing something "green".

The real burden of climate responsibility continues to fall upon those in the Global South.

Instead of focusing resources to move away from a fossil fuel economy, massive international trading ventures have been established as the solution.

But what impact are these offsets having? Are they actually reducing emissions?

THE CARBON MARKET

Due to the rising levels of carbon emissions around the world, a carbon market was established as one of the primary means to reduce pollution.

1. CARBON

1 TON = X $ = GOVERNMENT CREDITS

The carbon market, also known as emissions trading, works like a stock market and is set up to control pollution by providing economic incentives for achieving reductions in the emissions of pollutants.

A cap is placed on the total amount of emissions.

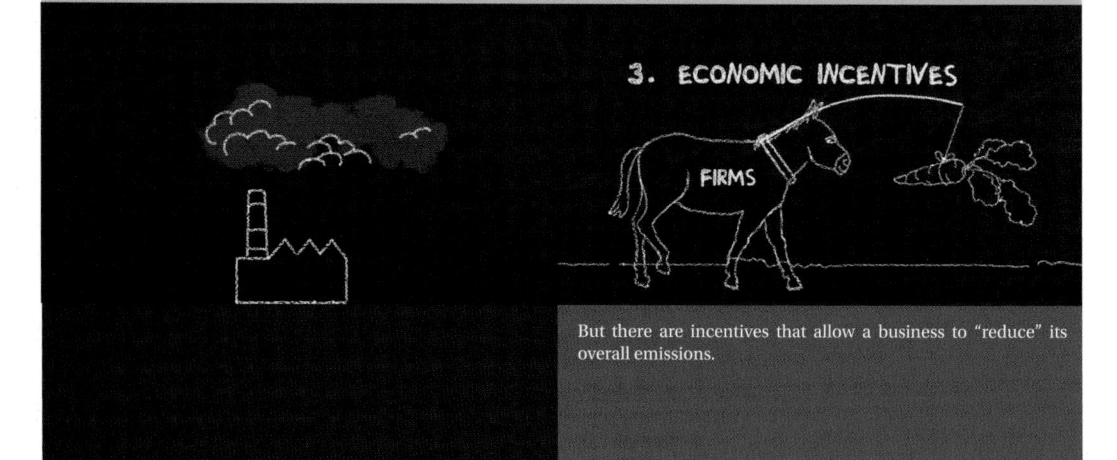

But there are incentives that allow a business to "reduce" its overall emissions.

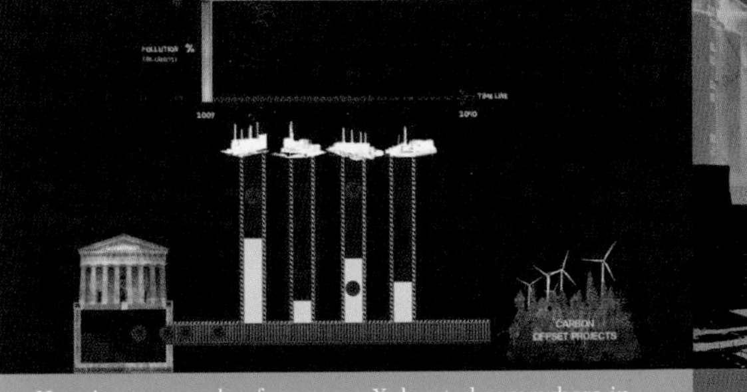

Here is an example of company X, located somewhere in the developed world.

The government allots company X's four factories each a set number of permits regulating how much carbon they can emit.

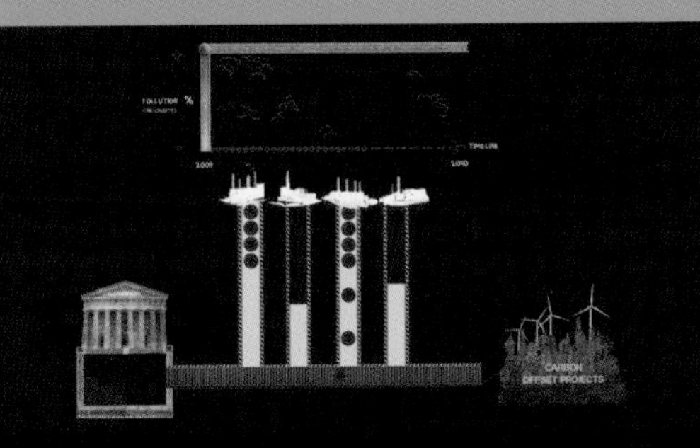

Factories producing less than the permitted amount of emissions can transfer their excess permits to factories producing too much.

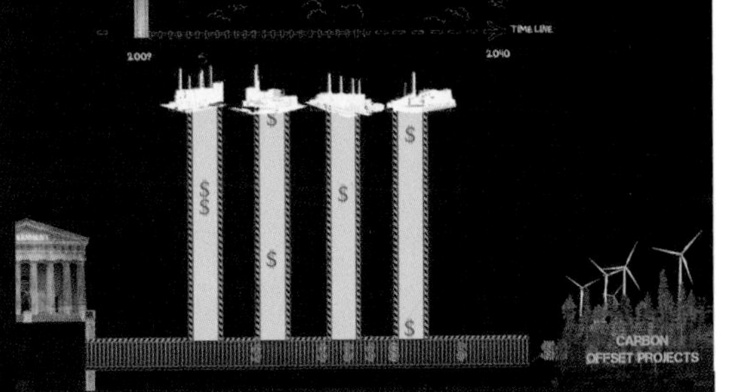

At this point, company X has fulfilled its commitment to reduce carbon emissions.

Meanwhile, big polluters make huge profits from this modern day Klondike.

It was, and is, the only International treaty to set hard limits on greenhouse gas emissions.

Kyoto set up three market mechanisms to assist industrialized countries to meet their targets for emissions reductions.

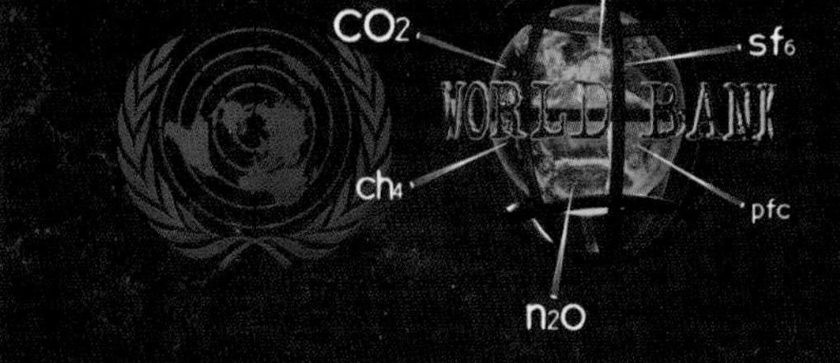

All three are different forms of carbon trading.
Since Kyoto's inception, the World Bank has expanded the role and infrastructure of carbon trading.

When the World Bank launched its first Prototype Carbon Fund in 1999, it was presented as a short-term catalyst to jump start the international carbon market.

The World Bank's portfolio has since grown to over 2.5 billion dollars, distributed across 13 different funds.

If things continue on their current course, the Kyoto Protocol will likely be allowed to expire and wealthy countries will replace the binding reductions with a voluntary commitment.

If they are successful, carbon trading will be cemented as the foundation of carbon emissions policy for at least the next decade.

Welcome to the Carbon Rush...

A VIEW FROM SCOTLAND

Living within the Glow of Grangemouth—How One Group and One Community Are Making a Difference in Scotland

by Norman Philip, Friends of the Earth

Grangemouth residents have long complained about high rates of asthma as well as the smells and noise coming from the plant.

Visitors to the industrial town of Grangemouth are often concerned by what they see and hear when they first approach the town. The petrochemical complex, its rising steam and bright flares, dominates the townscape. The visual result of all the industry can be menacing by day, and eerie by night as the illuminated factories bath the town in an absolutely spectacular glow. Yet local residents just take it in stride, as if the whole thing was just plain invisible.

Grangemouth is located within the urban central belt of Scotland, between Glasgow and Edinburgh. The town, established around a port on the River Forth, remains the second biggest port in Scotland. However, since the 1940s, the economy has shifted and now focuses primarily on the petrochemical industry. The current roll call of corporations includes some of the largest players in the industry, including BP, INEOS, Dow, and Syngenta. One street alone, known as Bo'ness Road, is home to three of the country's biggest carbon polluters. The Grangemouth Refinery is the only oil refinery in Scotland. It processes around 200,000 barrels of crude oil a day and produces almost all of Scotland's vehicle fuel. The refinery produce is used by factories around town to manufacture petrochemical, agrochemical, plastic, and gas products.

Polluting companies in Grangemouth and across the Western world are currently buying credits from offset projects around the globe in order to compensate for their pollution at home.

Grangemouth, which is one of Europe's largest oil refineries, emits sulfur dioxide, nitrogen dioxide, and small particulate matter into the air.

In the late 1990s, British Petroleum (BP), then owner of the Grangemouth Refinery, launched a major effort to "green" their image, in part by offsetting their carbon emissions through investments in projects that reduce the production of greenhouse gases.

The refinery on both sides of Bo'ness Road is owned by INEOS, the third largest polluter in Scotland. The refinery has a similarly harmful track record on a social level.

In 2002, Plantar, an iron foundry company with operations in Brazil, threatened to switch from burning charcoal to coal in order to increase its capacity.

Residents of Grangemouth have always been asked to balance the environmental impact of living alongside polluters with the promise of jobs and prosperity. As such, the community and industry grew in parallel. At its peak in the 1970s, most adult males in town either worked for the petrochemical industry or at the port. Job prosperity attracted an influx of workers and new housing estates sprang up all around. The balance appeared to be mutually beneficial, with well-paid local jobs and a corporate engagement ethic that supported social clubs for employees and their families.

For years people had always believed that the companies, whose employees lived locally, were concerned with the health and safety of the community. Health risks were therefore always at the back of residents' minds. People believed that, on balance, if there were associated health risks, someone would certainly do something. All factories in town always claimed to be within the safe emission limits set by environmental regulators. However, given the fact that there are over 10 factories in town, there should have been concern about the cumulative impact of the cocktail of gases being released over such a concentrated area.

The destructive environmental impact of the petrochemical industry was first brought to attention with the discovery of "acid rain." For perhaps the first time, people started to realize that pollution released in one country could poison the environment in another country. This new knowledge created the need for international agreements to address pollution. As a result, governments have made efforts since the 1970s to reduce the release of sulphur dioxide into the atmosphere, here as elsewhere, with varying positive results.

Over the past 15 years the names on the factories around Grangemouth have changed regularly with little fanfare.

Norman Philip from Friends of the Earth, Grangemouth, driving down Bo'ness Road explaining how the power plant in the community is the 6th biggest polluter in Scotland. On this one road, there are three of the top 10 biggest polluters in Scotland.

There is now less and less promise of local jobs. And those who now work in the industry usually live somewhere else and commute to town on a daily basis. The corporate social clubs have been closed down, boarded up, or demolished. The relationship between corporations and the local community is breaking down.

***

One of the first environmental projects I initiated in Grangemouth with Friends of the Earth was inspired by a visit from Denny Larson, executive director of the US-based Global Carbon Monitor that carries out campaigns with industrial communities fighting for justice around the world. The Friends of the Earth project addressed the apparent disinterest of local residents toward their industrial neighbors. So instead of trying to hold public meeting, which would have been for the most part ignored, we went onto the streets and asked residents, "When are you reminded that you live next to the petrochemical industry?" The feedback was written up in a report called "Living within the Glow—Stories from the Fenceline," which we hoped would be a catalyst for wider discussion within the community about environmental justice issues. As Grangemouth has so many carbon-intensive facilities in one concentrated area, the town makes a

On a longer term level, offset projects diminish human rights in several more indirect ways: First, boosting the fiscal success of the corporations participating in them, who facilitate any unsustainable practices a corporation might sponsor.

Second, an emphasis on greenhouse gas emissions has ignored the fact that offset projects sometimes produce highly toxic, non-greenhouse gas pollutants.

considerable contribution to Scotland's total carbon emissions. If carbon emissions could be cut in Grangemouth, there would be important benefits to the local, national, and global environment.

Within the Scottish context, the government has set challenging climate change legislation. The Climate Change (Scotland) Act requires an 80% reduction of all greenhouse gases by 2050, relative to 1990 levels. The Act also sets a limit on the use of offset credits to meet these targets, stating that a minimum 80% of the emission reduction effort must be made through domestic effort.

Unfortunately, big business has been slow to address its own carbon emissions and has looked at a number of so-called solutions, such as carbon offsetting, carbon trading, and technological fixes, which enable them to continue with business as usual while the levels of carbon in the atmosphere continue to rise toward catastrophic levels.

In 2005, Carbon Trade Watch, an independent research collective working on climate change and climate policy, were working with residents from the Espirito Santo community in Brazil who were impacted by eucalyptus plantations, cultivated as a carbon-trading prototype funded by the World Bank. People living next to the plantations were suffering because access to farmland was now restricted and water resources were being depleted as the eucalyptus sucked the water tables dry. Corporations like BP, who operate in Grangemouth, contributed to

Norman explaining how research making any statements regarding the impact of pollution on the local community has always been challenged by the industry.

the World Bank fund that had paid for the eucalyptus to be planted. Carbon Trade Watch wanted to link the community in Brazil with a community impacted by pollution in the UK where a corporation was engaged in the carbon-trade market. The "Living within the Glow" report was an introduction to the Grangemouth experiences.

Residents in the Brazilian community were shown how to use video equipment to make their own films and to share their experiences and feelings. We called these screenings "Video Letters from Brazil" which we then showed in Grangemouth. Issues from Brazil included the personal impact of water loss on families—from the loss of plants and herbs used for traditional medicine to the impact on family cheese making. Also included in these screenings was how the eucalyptus plantation divided the community, between those who were offered jobs and those who were not. And the people of Espirito Santo also talked about the consequences of speaking out against the plantation, which resulted in threats from business interests, and also from those on the other side of the divided community. All said, the videos highlighted the hidden human consequences of these carbon emission "solutions" and brought home to the people of Grangemouth the interconnected realities of the new pollution-trade economy.

After the Grangemouth screenings, local residents replied to the Brazilian videos expressing their support and solidarity, and sharing the Scottish experience of living next to a polluting industry. By exploring the Brazilian experience the residents of Grangemouth were able to understand their own experiences in a new light. By discussing the injustice in Brazil they were more able to consider the injustice in their own community. Altogether there was anger: anger that corporations in their own town could legally avoid a pollution reduction strategy, and anger that the process that allowed them to do this was negatively affecting people on the other side of the globe.

I believe that the carbon being released

INEOS has come under fire for its involvement in another carbon offset project with its own set of human rights violations....

in Grangemouth, which is released with a cocktail of toxic gases (including benzene, sulphur, and mercury) can and should only be reduced at source, right here in Scotland. The planet requires immediate reductions in carbon emissions to mitigate wide-scale climate change—and to make Grangemouth a more healthy place to live. Communities across the globe are suffering because, with the current carbon-trade system, it's business as usual, all the while climate change continues to spiral out of control.

The video exchanges between the Scottish and Brazilian communities became the documentary, *The Carbon Connection*, which was shown in Grangemouth before it was sent to Espirito Santo in Brazil. This film has subsequently been shown widely not only to expose the hidden impacts of carbon trading, but to show that local voices can be heard and can find an international audience. The voices of concerned people do have influence.

A recent investigation by the UK's Daily Mail found "dangerously high levels of fluoride and chloride-fluoride in the water was more than twice the international acceptable limit. All the water fell well below any safe drinking standards and the soil had worryingly high levels of these chemicals."

***

In 2005 INEOS, the world's third largest chemical company, bought the Grangemouth Oil Refinery from BP. That year INEOS was involved in a carbon offset project with Gujarat Fluorochemicals Limited (GFL), a company in Gujarat, India that produces HCFC-22, a refrigerant gas for air conditioning units and refrigerators. INEOS supplied the technology to capture and recycle HFC-23. This potent greenhouse gas is a by-product of GFL's operations. Both companies received the rights to claim the carbon reductions through the Clean Development Mechanism (CDM). The Gujarat project awarded INEOS and GFL an undisclosed number of Certified Emission Reduction units. What the CDM did not take into account was the environmental and health impacts on the Indian community due to the dangerously high levels of fluoride and chloride in local water and soil. By only concentrating on carbon reductions, both INEOS and GFL were able to continue to pollute in communities in Scotland and now, through support of the CDM, in India as well.

Today, Grangemouth residents are becoming more aware and more vocal. Currently, they are concerned about the latest promise they have been given by big business — low carbon sustainable development within their town. The local port operator in partnership with a power company is proposing to develop a Biofuel Energy Plant within 200 meters of existing homes. Residents are once again concerned that they are being asked to take on more than their fair share of pollution in the name of the national good. But residents are now more globally aware and have a wider understanding of the environmental justice implications of the proposed development. In fact, residents have voiced their concern that the fuel stock for the plant will be exploited from forests across the globe which might carry subsequent environmental and human rights implications as well. The claim that burning biofuels can be carbon-

1. But these chemicals do not contribute to global warming. Thus, those monitoring the program considered it successful insofar as it reduced greenhouse gas emissions.

2. The CDM Executive Board approved the Gujarat project in 2005 and awarded INEOS and GFL an undisclosed number of Certified Emission Reduction (CERs) units over time (INEOS' website predicts that together with a second similar project in Korea, the Gujarat project will generate three million tons of CERs annually). Both companies were then free to sell those credits to industries falling short of their national emissions caps.

3. For instance, in 2006, GFL made news for doubling its sales revenue by selling a record number of carbon credits to Noble Carbon Credits group of Singapore, Rabobank Nederlands and Sumitomo Corporation, most of which resold the credits mainly to large industries needing to comply with Kyoto Protocol, including power and oil companies.

4. According to some reports, some of the proceeds from GFL's sales went to build a Teflon and caustic soda manufacturing facility which uses processes known to be massively polluting.

Offsets have allowed the continued pollution of the Grangemouth community, and they introduced new hardships for people in Brazil. More indirectly, the notion that Grangemouth's emissions were being neutralized made it an attractive asset that increased the profitability of its various owners, enabling them to invest in other toxic projects.

neutral is no consolation if you live next to the smoke stack.

Laws, legislation, and business models that focus purely on carbon have allowed corporations to turn pollution into a commodity. This lack of foresight has contributed to environmental injustice within communities across the globe and delayed the real actions required to tackle climate change. Through awareness and solidarity we need to challenge the "business as usual" model and demand the urgent action required today.

No community should be asked to take on an unequal share of pollution caused by business, and no corporation should be allowed to abuse human rights in the pursuit of profit. When personal experiences are connected to voices from other communities around the globe, and to wider environmental issues, we are empowered to act to effect change on the problems that affect us.

FOREST SECTOR NEW DYNAMICS

Carbon Credits and CDM Projects—New Strategies and Grand Problems in Brazil

by Tathiane Paraíso Silva
Germana Platão Rocha

It was the savannah that provided fruit, food and space for cattle. It provided water for the areas where people could get food. It was destroyed by the arrival of large-scale eucalyptus monoculture plantations in the region."

The forestry sector is increasingly recognizing that good environmental management is good for business. As such it is greening up with carbon credits and Clean Development Mechanism (CDM) projects to gain "green" recognition that will help increase new investment in new markets.

Profits aside, the interests of the forestry sector have severally impacted on local communities who live on and near CDM projects, especially here in Brazil.

Discussions about climate change and the use of so-called CDM projects to mitigate it have started to intensify. And such talks will become even more urgent as industrialized consumption trends increase and the irreversible impact on the environment continues. The magnitude of this disaster is scary and a real cause of concern for both society and the scientific community.

However, many international treaties and conferences aimed at lessening climate variations are not achieving their goals. Truth be told, many end up intensifying pre-existing problems and at the same time conjure up all kinds of new social and economic issues. The Kyoto Protocol, signed in 1997, is one such agreement. Its goals are to reduce gases that cause global warming, mainly from developed countries, but criticism about its implementation has focused mostly on deadlines and targets. We

Members of the Casa Brava community commemorati the death of Antonio Joaquim.

The physical death of Antonio Joaquim adds another to all the deaths in the years since the arrival of eucalyptus farming to the region.

need, however, to look more closely.

We realize that the Protocol has failed to achieve some of its goals, especially concerning countries that pollute most. But as we know, the Protocol has also generated a trade market on carbon emissions through something called "carbon credits." Carbon emissions have thus become a currency to barter with, not reduce. Companies from developing countries are not necessarily obliged to reduce emissions, so they can trade credits with other countries as a kind of compensation. Will this solve the climate change problem? We don't think so. Global warming is real and it has affected human essentials such as food and access to clean water. It has even impacted on Western production and consumption patterns. Maybe this is the real reason climate change has reached such a high level of concern today.

The CDM is the only process that includes developing countries in the Protocol. Developing countries, those in the South in most instances, are seen to have clean and renewable energy resources that Northern nations can invest in to compensate for their carbon emissions. The result is not reduced overall emissions, but instead, the development of "sustainable projects" throughout South and Central America (and elsewhere) that are wreaking havoc with local populations and the environment.

The process of buying and selling carbon credits occurs through negotiations with developing countries that have not yet reached the need to reduce their emissions as set out in Protocol quotas. (By convention, a ton of carbon dioxide [CO_2] corresponds to one carbon credit.) In this way, developing countries can sell their quotas to those countries that are not yet able or willing to reduce the levels of pollution they release in the atmosphere. According to the BM&FBOVESPA[1], each ton of CO_2 equivalent (tCO_2) not emitted or removed from the atmosphere by a

1 The BM&FBOVESPA, a Brazilian-owned company formed in 2008 from the integration of the Bolsa de Valores de Sao Paulo and the Mercantile & Futures Exchange. As the main Brazilian institution of brokerage operations for the capital market, the company develops, implements and provides systems for equities, equity derivatives, fixed income securities, government securities, financial derivatives,spot currencies and agricultural commodities. Available at: http://www.bmfbovespa.com.br.

1-2. The Vallourec and Mannesman eucalyptus plantations are part of the Vallourec project on renewable energy registered as a carbon offset since 2006. Vallourec claims to be using "renewable" energy for pig iron production, with the supposedly renewable energy coming from their eucalyptus plantations.

3. With the arrival of eucalyptus monoculture, a local source of sustenance, was destroyed. Here, cattle look for pasture under the ecalyptos.

In Brazil there are many eucalyptus monocrop projects operating by numerous different companies in various regions. They all have similar impacts on the local communities.

developing country can be traded on the world market. This is the current state of global emissions reduction.

Although these agreements are settled, societies around the world still have to deal with the resultant and ongoing green initiatives, absorbing the whole potential damage that they might cause. The go-green speeches often change but the practices remain the same: that is, the climate crisis at hand has become another source of profit for large corporations, including the forestry sector.

Given this, large "reforestation" companies along with modern cattle ranches compete today for space with diverse communities that sustain themselves on traditional uses of natural resources. This immediately creates conflict, especially in the Minas Gerais, Brazil, which is a state with the largest area of continuous eucalyptus plantations in the world. According to the Association of Forestry of the State of Minas Gerais, forest plantations in Brazil occupy an area of approximately 5 million hectares, equivalent to 0.6% of the country. The area of planted forests in Minas Gerais reached an incredible 1.4 million hectares of planted eucalyptus forests in 2010.

Minas Gerais had all the favorable conditions required by the eucalyptus industry: it was a state full of lands considered "empty and low valued," and it was full of raw materials and skilled labor that contributed to the Brazilian dependence on steel production. The North Minas Gerais region was seen by policy makers as not very attractive to industry. To integrate this area into the development of the country, the State provided favorable conditions for the

The way the charcoal production works is that the companies first clear the vegetation to plant the eucalyptus.

1. The companies hire some of the local work force to work on a temporary contract basis.

2. When it is time for the harvest, the company then hires a handful of workers to cut the eucalyptus and then transport it to the ovens.

3. The temporary local workers also feed the ovens which burn the eucalyptus to make carbon rods.

4. Once the carbonization process is complete, the workers go back to empty the ovens..

reforestation companies to settle in the region, through incentives such as tax exemptions. Thus, the northern portion of the State became integrated into the economic cycle of Minas Gerais, providing sufficient resources to supply the steel mills in the central region.

As the eucalyptus was brought to the cerrado (Brazil's tropical savannah), deforestation was intensified. Native forests were carbonized (turned into charcoal) for use by industrial consumers. All this played out in other regions but it is now brought here to the Minas Gerais and subsidized through carbon offset project.

The growth trajectory of silviculture has left Minas Gerais with the highest production of charcoal from planted forests, something exceeding more than 78% of national production. This growth, according to Bacha and Barros (2004), was primarily due to increases in planted forests made in 2000 and 2001 by pulp mills and steel mills that benefit from the charcoal.

The need to increase the scale of eucalyptus production, and additionally the exploration and concentration of land holding, triggered the expropriation of small farms. Proud, independent landowners became employees. In other words, plantations started to replace whole groups of farmers and their family economies. They were then given only subsistence wages that would not secure more than their own survival.

***

The Cana Brava community, located in the city of Guaraciama, is one of several

After that the charcoal rods are transported to the metallurgical sector of Minas Gerais.

Brazilian communities directly affected by eucalyptus. The community is comprised of small farmers and, due to the projects and incentives that are "modernizing" the countryside, find themselves trapped by the monoculture of eucalyptus and excluded from public policies that would ensure their stability as a community—and their safety. In 2007 for instance, a farmer was murdered by one of the security agents from V & M Florestal, a company that acts in the region. The man was accused of stealing wood from land said to belong to the company. So far, the incident has never been brought to trial. This fact generated national and international interest, where several non-government organizations and social movements have provided support to the community, and are especially addressing the political education of residents.

The eucalyptus company has caused many economical, environmental, and especially social changes to the community. With privatization of land came the possibility of rent. Traditional farmers were termed squatters and expelled and the land then turned over to plantations. Lives were destroyed.

These spaces were important to the community as a means of survival for the population living there. The cerrado represented more than the environment of these people; it was a source of income and livelihood of the population, and an essential element of the reproductive system of the community.

***

The rods are transformed into pig iron. It is also used for steel and sheet metal. Much of the steel and sheet metal is exported outside of Brazil to be processed into cars and then returned in its final product.

The elimination of native forest and traditional agricultural land and its replacement with eucalyptus monoculture has become much more than a simple environmental issue. It has fuelled a profound change in the habits and manners of the local community, and it has affected other issues such as income, social and family bonds, and even changes in eating habits.

Despite all the problems that were caused, community members are now concerned most with carbon-offsetting incentives that will further entrench the monoculture and its consequences in the area, because these new incentives will cause further damage to people who live daily with the problems already caused by climate change "solutions."

That is why we must look more closely at the new practices and mechanisms that express support and encouragement to these "clean technologies" that carry big price tags and sweeping consequences. And we must first take into account how these projects will affect the people and communities surrounding them. Blind and reckless economical development should by now make us rethink how and who should manage the forests and people who inhabit them. Further, recognition and visibility should be given to these often spatially segregated communities, giving value to the knowledge of these populations, and recognizing their essential contributions in the maintenance and preservation of natural resources and traditions.

1. During its life cycle, eucalyptus captures carbon dioxide from the atmosphere. If it were to remain in nature then there would be no contradiction. However as it is cut, burnt and turned into pig iron, releasing carbon dioxide back into the atmosphere, it is difficult to understand how this is reducing carbon emissions.

2. Dona Zeza explains how before the eucalyptus plantations were introduced there were many fruits that the community could harvest but since the eucalyptus plantations there is regular water shortage. In other words, the eucalyptus sucks up all the water that should be used for farming, making it very difficult for the local community to survive.

3. The people of Cana Brava are regularly harassed by the company guards who are patrolling the only access roads to the community.

4. Is hope dawning on Casa Brava?

REFERENCES:

Acselrad, H. **Environmental policies and democratic construction**. In: Viana, G., Silva, M., Diniz, N. (Ed.) The challenge of sustainability: a socio-environmental debate in Brazil. Oxford: Foundation Abramo, P. 75-96. 2001.

AMS. **The forest industrial complex in Minas Gerais (CIF): characterization, measurement and importance**. Belo Horizonte, in October 2004.

BARBOSA, RS & FEITOSA, AM. **The dynamics of the struggle for land in northern Minas Gerais.** In J. Cleps Jr., JA Zuba, AM Feitosa (Eds.). (2005). "Under the canvas: regional trends and challenges of the struggle for land and agrarian reform in Brazil." Goiânia-GO. Publisher of the SCU.

Chévez POZO, OV **Property regimes and natural resources: the tragedy of the privatization of common resources in the north of Minas Gerais.** Rio de Janeiro: Doctoral Thesis. UFRRJ, 2002.

Goncalves, Múcio Tosta. **We timber: social change and employees of the forest plantations in the Steel Valley / Rio Doce in Minas Gerais**. Doctoral Thesis presented to CPDA-UFRRJ: Rio de Janeiro, 2001.

Goncalves, Múcio Tosta. **Farmers and Workers of Forest Plantations in Minas Gerais: What are the problems¿** Paper presented at the X Seminar on the State economy. Diamantina, MG, 2002.

TRASHING OUR FUTURE

False Solutions to Waste Management

by Dharmesh Shah
India Coordinator for Global Alliance
for Incinerator Alternatives

"In this century of progress, with our knowledge of chemistry, and with the most complete machinery at our disposal, it seems to me like a lapse into barbarism to destroy this most valuable material simply for the purpose of getting rid of it, while at the same time we are eager to obtain these very same materials for our fields by purchase from other sources."

Chemist Bruno Terne, 1893

The Okhla site, in South Delhi, will also burn this Refuse Derived Fuel (RDF) to generate electricity.

As our civilization continues to grow rapidly, we seldom spend time considering the repercussions of this "progress." Fact is, the biggest casualty of this growth has been our shared environment. A lot of what we do on this planet has impacted on it in different ways, but no activity has had as deep an impact as waste. Garbage, refuse, waste—this is the undesirable reality of modern civilization that fills our streets, chokes our water bodies, and pollutes our oceans.

Waste today is an international problem and cannot be looked at in isolation. In our globalized economy, raw materials are constantly extracted, processed in factories, shipped around the world and ultimately, burned or buried in communities at an ever increasing rate. Yet solving this problem requires much more than technical fixes: how we manage waste is part of a larger web of decisions about health, equity, race, power, gender, poverty, and governance[1].

In countries like India and China which lack the resources to tuck it out of sight in landfills, waste is a stark reality of everyday life. Open garbage dumps are an accepted reality in all these urban landscapes, often strategically banished to the municipal limits of a city. Here, on the outskirts, they become a burden to be borne by

1. Zero Waste for Zero Warming— GAIA's Statement of Concern on Waste and Climate Change, 2008.

The Timarpur-Okhla Waste Management Company project was registered by the Clean Development Mechanism to build two facilities to handle more than 2,000 tons of municipal waste per day.

marginalized communities or specifically, in the case of India, by those ousted from its traditional caste system.

The areas surrounding such dumps are perpetually surrounded by a veil of toxic smoke belching out from the smoldering garbage while the smell of putrefying food hugs the air—not to mention the thousands of tons of methane seeping out into the air. Yet one can see thousands of people living around these dumps and hundreds rummaging through the garbage piles eking out a living on what the city threw out.

On a social and ecological level, the implications of such waste-racism are far-reaching. If we connect the dots, our collective mismanagement of waste can be linked to all other environmental crises facing us today—from urbanization and poverty to resource conflicts, deforestation, and climate change.

False Solutions

Clearly, there are serious limitations to the ways we have adopted to live our lives. And it has now become evident that our mismanagement of the subsequent waste is one of the reasons behind the environmental crisis we face. But alas, solutions that have emerged in response to this garbage crisis have merely been created to generate profits and offer nothing in terms of sustainability.

There are nearly 1,800 waste management companies[2] worldwide offering treatment, processing, disposal, and incineration services. According to the World Bank, urban areas in Asia now spend about $25 billion US on solid waste management per year; and this figure will increase to at least $50 billion in 2025[3]. With such wind-

2. Green Pages Directory: http://www.eco-web.com/reg/sfc.html?p=06667.html

3. What a Waste— Solid Waste Management in Asia. The World Bank, May 1999.

Both facilities are meant to produce Refuse Derived Fuel, otherwise known as RDF.

fall profits on offer, several companies are making a beeline for Asian cities looking to outsource their waste management. Traditionally, waste management has been the responsibility of Urban Local Bodies (ULBs). In India, rapid urbanization, poor planning, and unprecedented increases in waste generation rates have coupled with the failure of local municipalities to move away from centralized dumping. These factors have led to a dizzying garbage problem that now serves as an ideal entry point for big, private waste management enterprises to present "complete solutions" to cities struggling with waste emergencies. On the table are large-scale projects like incinerators, landfills, Refuse Derived Fuel plants, and mixed waste composting facilities costing hundreds of millions in tax-payer dollars.

As we'll see, such projects not only aggravate problems resulting in the destruction of valuable resources, they also undermine recycling patterns and directly compete with real solutions. Regrettably, these big waste management investments are also the most attractive for governments and policy makers looking for easy packaged options.

Incinerators and the Environment

Most dangerous of these are incinerators. Incinerators give our environment its major dose of dioxins and furans, highly toxic man-made substances. Dioxins and furans are listed among the "dirty dozen"[4] chemicals slotted for elimination under the Stockholm Convention on Persistent Organic Pollutants (POPs) due to its severe impacts on the environment. Dioxins and furans cause a wide range of health problems, including cancer, immune system damage, reproductive

4. The Dirty Dozen—Persistent Organic Pollutants: A Global Issue, A Global Response. USEPA, 2009.

Anita Wadavadekai, a waste picker, explains, "We used to pick and sell the sorted waste at 10 rupees ($0.25) per kilo. I've been doing this work since my childhood. We were picking solid waste. When the SWaCH cooperative explained what good we were doing by picking waste, we understood that we were helping the environment, the planet. Now, we don't even earn 100 rupees ($2.50) per day. But we didn't know."

1. Rebecca Kedar, a waste collector, explains the impact of Henjer receiving the waste to energy contract, "Because of the Hanjer company, 2,000 to 3,000 people from the SWaCH coop have lost their livelihoods. Whatever few rupees we used to earn have stopped, and the Hanjer Company is responsible for this loss."

2. Shahid Hasan, a resident of Delhi, explains, "What you see on the right hand side is the 2,000 ton waste incinerator coming up.This is barely 150 metres from this residential neighbourhood. This project started, the construction started in 2010 and it's almost near completion.The residents of this neighborhood and other neighborhoods have been protesting for the past three years about this project. But all this time, no one has listened to our memorandums."

3. Ravi, Agarwal Director of Toxics Link, explains how "Carbon markets are becoming one major revenue stream which is tipping the financial equation in favour of these projects. If you take this revenue stream away, these projects lose their traction for most private parties to want to invest in. All of the stakeholders who are promoting this project are marketing this in the name of the climate change business, that this is entitled to getting carbon credits and they are tagging it as "green" energy or as a "green" project. And they try to confuse the people here. Those who are well aware of the concept understand the process of waste incineration. None would agree that this project is safe for the people who live close to this project. When you try to control emissions in these machines it is really expensive. A good sized incinerator, for example, that burns 2,000 tons a day in Europe, can cost up to $500 million. In a country like India, and other such countries, it's almost impossible to make that kind of investment. So what you do is cut corners. You make burners that are not up to the mark, which will not have all the pollution control equipment. They will just release all kinds of chemicals into the air.

and developmental disorders. They also bio-magnify, meaning they are passed up the food chain from prey to predator, concentrating in meat and dairy products and, ultimately, in humans. They are of particular concern because they are ubiquitous in the environment (and in humans) at levels that have been shown to cause health problems, implying that entire populations are now suffering the ill effects. Worldwide, incinerators are one of the primary sources of dioxins and furans.

Incinerators are also a major source of heavy metal pollution. Among them is mercury, which is a powerful neurotoxin known to impair motor, sensory, and cognitive functions. At highest risk are children. Other heavy metal pollutants such as lead, cadmium, arsenic, and chromium cause a whole range of health problems.

Additional pollutants of concern include other (non-dioxin) halogenated hydrocarbons; acid gases that are precursors of acid rain; particulates which impair lung function; and greenhouse gases. However, a full characterization of incinerator pollutant releases is still incomplete, and many unidentified compounds are present in air emissions and ashes[5].

The environmental risks of incineration gave rise to one of the most diverse global movements in the 1980s and 90s. Most of these struggles were in developed countries where prosperous local governments embraced incinerators in

5. Waste Incineration: A Dying Technology. GAIA, 2003.

The Henjer RDF Pune plant is currently on the consideration list of the United Nations to receive carbon credits as a reusable energy source.

The Timarpur-Okhla project will reduce emissions by an average of 26,000 tons of CO_2 equivalent per year.

their bid to "modernize." In the United States, for instance, business interests and a perceived landfill crisis drove an incinerator-building boom during this period. But this boom spawned a massive grassroots movement that defeated more than 300 municipal waste incinerator proposals. The activists fought for higher emission standards and removal of subsidies, which virtually shut down the industry by the end of the 1990s. Similarly, in Japan, the most incinerator-intensive country on Earth, resistance to incineration is nearly universal, with hundreds of anti-dioxin groups operating nationwide. Public pressure there has resulted in over 500 incinerators being shut in recent years. In Europe, resistance has taken the form of alternative action. Some regions have cut waste generation dramatically even as populations have climbed. As a result, there is little market for new incinerators in Europe[6].

Incinerators in Disguise

Widespread public disapproval led to the death of the incineration industry—but business interests in waste persisted. That's why incinerators have made a pseudonymous comeback in the guise of waste-to-energy (WtE), pyrolysis, and plasma arc gasification schemes. These new-age "incinerators in disguise" came with an attractive option: they would extract energy using some of the heat generated during the combustion process. This energy alternative gave the industry a new lease on life. However, all the threats associated with traditional incinerators still held true for these technologies.

Communities worldwide are now facing an unprecedented onslaught of proposals from waste treatment companies and entrepreneurs promoting new-generation incinerators. These companies claim they can safely, cost-effectively, and sustainably turn any type of garbage, such as household trash, tires, medical waste, biomass, and hazardous waste into electricity or into fuels like ethanol and bio-diesel[7]. Some companies go so far as to claim that their technologies are "zero emission" or "pollution-free" and not,

6. Waste Incineration: A Dying Technology. GAIA, 2003.

7. Incinerators in Disguise—Case Studies of Gasification, Pyrolysis, and Plasma in Europe, Asia, and the United States, GAIA and Greenaction for Health and Environmental Justice, April 2006.

in fact, incineration at all. However, all these technologies emit dioxins and other harmful pollutants into the air, soil, and water. And they are defined as incineration by the US Environmental Protection Agency and the European Union[8].

Incineration in India

Developing countries in Asia such as India and China provide fertile markets for this emerging industry. Urban consumption patterns in these countries are fast mimicking those in the West, but generally, the composition of waste is largely organic with comparatively little plastic and paper. These factors render the waste useless for combustion. Nevertheless, this has not deterred waste management companies from making inroads into municipalities eager to "modernize" their waste management. There have also been several documented cases of fraud, collusion, and misinformation by project proponents desperate to make quick profits by peddling untested technologies upon these cities.

India's first modern waste-to-energy (WtE) incinerator, proposed at Okhla in New Delhi, was under construction during the shooting of *The Carbon Rush*. The project was built in the face of intense protest from local residents concerned about public health, and waste picker unions concerned about how the incinerator might destroy the livelihoods of their members. Set up in the midst of densely populated South Delhi, human settlements and hospitals are barely 100 meters from the proposed site. The project claims to have the best pollution control technology but with an investment of merely $40 million US, it cannot afford the best technology. In comparison, a refuse derived fuel (RDF) incinerator used for processing 230,000 tons per year which incorporates the best German technology was set up in Rostock, Germany at a cost of 83 million Euros ($110 million US).[9]

The other impacted section, the waste pickers of Delhi, have been pitched in a non-legal battle against the project since it was first announced. Several demonstrations and rallies have tried to draw some attention to the plight of the nearly 100,000 Delhi waste pickers whose livelihoods will be set to fire, literally. Numerous studies around the world and in Delhi underline the invaluable contribution of this informal sector in waste recovery. Waste management models that integrate waste pickers have numerous positive social, financial, and environmental benefits which have also been well documented[10]. According to a study on the informal waste sector in Pune, India done by the International Labor Organization (ILO), each waste picker contributes 246 rupees worth of free labor per month by recovering the waste. This amounts to a saving of over 89 million rupees ($1.8 million US) for the municipal corporation[11] that would have otherwise been spent on waste transportation and landfilling. Communities and waste pickers across India are now pitched in similar battles against like proposals in cities such as Mumbai, Pune, Calcutta, Chennai, Bangalore, and Trivandrum.

The case in India also applies to most Asian countries as local governments across the continent are diverting large

8. U.S. Environmental Protection Agency, Title 40: Protection of Environment, Hazardous Waste Management System: General, subpart B-definitions, 260.10, current as of February 5, 2008.

9. List of incinerators in Germany: http://www.industcards.com/wte-germany.htm

10. Timarpur-Okhla Waste to Energy Venture by Dharmesh Shah, GAIA, 2011.

11. ILO Study of Scrap Collectors, Scrap Traders and Recycling Enterprises in Pune 2001.

By comparison, Delhi's waste pickers will contribute to a CO_2 reduction 9 times more than Timarpur-Okhla project can do.

The project revenue for the incinerator is expected to be 37 million dollars from the sale of carbon credits alone.

chunks of their investments into WtE incinerators. Incinerators are so expensive they often take up all capital available for waste management. Furthermore, after building an incinerator, municipalities often become reluctant to allocate funds to recycling and composting programs that can reduce the quantity of material available for incineration.

Clean Development Mechanism (CDM) and Waste Management

The Clean Development Mechanism (CDM) targets the waste management sector as an area where greenhouse gas (GHG) emissions can be reduced. While the revival of the incineration industry certainly has a lot to do with the way it has repackaged itself as a generator of energy, no other body has played as perverse a role as the CDM in its revival. The CDM made incinerators eligible for participation in the carbon market thus endorsing the industry's false claims of being green.

Ironically, by promoting incinerators the United Nations Framework Convention on Climate Change is working against the mandate of the UNEP's Stockholm Convention on Persistent Organic Pollutants (POPs).

More than 1,000 CDM projects worldwide have qualified for carbon credits. Most of these are large-scale activities in the energy sector. In the waste sector, subsidized technologies include landfill gas, incineration, and cement kilns which can co-incinerate waste. India and China are the biggest generators of carbon-credits, with a combined share of more than 50% of the projects. With some 3,000 more projects awaiting registration, the CDM expects to generate nearly 3 billion Certified Emission Reductions (CERs) by 2012, when the first Kyoto commitment period ends[12].

Unfortunately, CDM has funded a large number of waste disposal projects like incinerators, landfill gas, and mixed-waste composting. As of May 2008, out of 90 projects funded by the CDM to improve municipal waste management, 83 were landfills with gas recovery, and another 5 included incinerators.

Most of these projects claim carbon credits on the basis of reducing methane emissions that would have otherwise been released from landfills and open dumps. But no consideration has been given to emission reductions currently made by better options such as recycling. For instance, it is estimated that the project in Okhla, New Delhi will reduce nearly 602,610 tons of CO_2 over 10 years from the time it commences operations. In comparison, the informal waste sector in Delhi—which supports over 100,000 people—currently reduces emissions by an estimated 962,133 tons of CO_2 equivalent each year by recycling glass, metals, plastics, and paper alone[13]. At the time of shooting *The Carbon Rush*, the Okhla project was under construction to be followed by two more proposals in East and North West Delhi which together aim at burning all of Delhi's waste. As a result,

12. What is Clean Development Mechanism? GAIA, 2009.

13. Cooling Agents — An Analysis of Climate Change Mitigation by the Informal Recycling Sector in India, 2009.

all of the emission reductions achieved by the waste pickers will be lost because these incinerators will burn recyclable materials. Hence, there will be an overall net increase in emissions.

In reality, incinerators contribute significantly to climate change, not only by releasing greenhouse gas themselves, but by wasting resources that can be reused, recycled, and composted, thus leading to increased raw materials extraction, and the manufacture and transport of new products—all of which are extremely energy-intensive processes. Despite this, the CDM qualifies incinerators for carbon incentives which indirectly undermines all real efforts.

Waste is yet another sector where CDM incentives to private companies and toxic technologies are pushing our environment and society away from sustainability. Such projects fail to address the root causes of the problem and therefore do not offer meaningful solutions. If the CDM aims to fulfill its mandate, it must cease wasting money on such projects and turn toward genuine solutions.

The Real Solutions

So, what do we do with waste? Over the last century, billions of tax payers' dollars have been spent on increasingly sophisticated technologies with the hopes of making waste disappear. But neither incinerators nor landfills truly dispose of waste; each creates significant, hazardous byproducts and generates additional waste streams that require further management. Our attitude toward waste is one devoid of connections to other aspects of our system. As long as we isolate waste with the view to "manage" it there is going to be little success.

We need to rethink waste and talk about resource management. We need to work toward creating a closed loop economy that can take discarded material that is of no further use to the present owner and feed that material back into the economy as a resource. This will ensure an end to the mixed waste stream. When discards are mixed, they become useless and people think they need large-scale disposal technologies to manage them. If discards are segregated at source, they are compatible with more sensible and effective management options such as recycling and composting.

THE WINDMILLS OF MAHARASHTRA

Mortgaging Life of Marginalized Communities in India

by Nashant Mate
Maharashtra Association of Social Work Educators (MASWE)

Since 2007 hundreds of different wind power offset projects have been registered with the UN's Clean Development Mechanism.

India has some 3,000 Clean Development Mechanism (CDM) projects either registered or at some stage of registration with the United Nations Framework Convention on Climate Change (UNFCCC). In all, it is claimed that these projects will reduce hundreds of thousands of tons of CO_2 in 2012 with estimates reaching into the millions by 2020.

CDM projects are clearly big business in India and they have popped up throughout the country at an accelerated pace. In Maharashtra for instance, a state occupying the central-west portion of the country, there are 231 CDM projects. Of these, 104 are wind energy projects. Altogether, Maharashtra has already been issued 2,246 Certified Emission Reductions (CERs), of which 980 CERs are from windmill operations.

Chalkewadi—The Saga of Broken Promises

Chalkewadi is set in the District of Satara, Maharastra State on the Western Ghats. Chalkewadi has a population of 1,200 whose basic means of livelihood depends on agriculture. The Maharashtra Energy Development Agency (MEDA) started a "Demonstration Windmill Project" at Chalkewadi village in 1996. Initially, the government leased 100 acres of land, which was not being used for agricultural

Today the Satara Region has more than 1,000 wind energy generators owned by Suzlon, Tata Motors and others on what was previously small-scale farmland.

Hajirao Rawali, a resident of Satara district, explains, "This area is all ours. They did not ask us and instead made a direct deal with Shahu Bhosale, a regional notary. Shahu Bhosale should not have dealt directly with Suzlon. If they had asked us, we would not have agreed."

purposes, from villagers on a 5-year term. The government later purchased the land at 6,000 rupees ($120 US) per acre. MEDA set up four power plants at the site, with installed capacities of 3.7 MW each. The apparent success of the project attracted private companies such as Suzlon Energy, a significant supplier and manufacturer of wind turbines and related equipment. Suzlon erected wind power plants in the area on the invitation of MEDA, and purchased lands within a 20-kilometre radius of Chalkewadi and other adjacent villages like Vankuswadi, Absari, and Kati. Land cost was 40,000 to 60,000 rupees ($800-$1,200 US) per acre.

Many companies like Bajaj Auto, Tata, Star, GIO, Sarita Chemicals, Westaj RRB, Energy Micon, and MTL later purchased these Suzlon power plants. Suzlon had already invested a huge amount—400 billion rupees ($8 billion US)—in the wind-energy plants and now, the valley has approximately 1,000 such plants. Of these, MEDA has 11, Suzlon has 67, Bajaj Auto has 100, and Tata Motors 76.

The villagers of Chalkewadi initially supported MEDA's Demonstration Windmill Project with great expectations that new employment opportunities and other developments would come to their village as well. Ramachandra Tatyaba Chalke, who has been the Sarpanch (elected leader) of the village for the last 26 years, provided his support. And so keen were the villagers that they even gave shramdaan (voluntary labor) for one week to construct the approach road to the village.

During construction of the MEDA and the first Suzlon plants, the villagers earned about 100 rupees (about $2 US) a day in wages. But now, such wage work is rarely available because of the increasingly specialized and technical nature of jobs. The plants employ 46 people from Chalkewadi as permanent staff. Only a few others get employed on a contractual basis. Chalkewadi's Gram Panchayat, the local self government, receives 2,500 rupees ($50 US) per plant as a form of annual levy from Suzlon. The levy is collected for 70 plants in the Panchayat area, which villagers use in developmental activities.

In 2003, people from four talukas (local administrative districts) organized a protest against further construction

of wind turbines in the area because they believed the plants drove away the clouds, causing decreases in rainfall. People from all talukas including Bid, Patan, Mann, and Satara suffered from unprecedented drought for three years in succession. Though the prolonged and intense monsoon in 2004 took the edge off the agitation, people were still very wary about the turbines.

It is known that representatives of Suzlon regularly meet villagers to discuss the problems caused by the plants. In spite of that, there is complete disengagement between local villagers and the company. As the villagers have sold their lands, they are indifferent to the company's affairs and, according to many, meetings with Suzlon are routine affairs and nobody is much interested. It is like they don't care anymore and have given up.

Bhambe Village

Bhambe is located in the Sahyadri Valley, Western Ghats. The ongoing wind energy project here is spread over a number of valleys. The river Koyna flows through one part of the project area and is bordered by thick forests. The area had already been announced as a "sanctuary," cutting people's access to it. To compound this, a dam named Tarali has also been constructed on the river which has displaced many families. Bhambe is located very near to this dam. Here, Ellora Time Ltd. acquired about 250 acres of land from villagers and 70 wind turbines were subsequently erected within the area. These wind turbines stand on agricultural lands held by private farmers as well as village commons under control of the Gram Panchayat. Helped by political leaders, the company

did not have to come to the village to finalize the land acquisition deal; they did it in the city. Prior to construction of the windmills, Bhambe villagers had no knowledge that their lands were being handed over to the company.

Not surprisingly, villagers tenaciously opposed the new construction. So the company sought help from the local administration, and the entire community of Bhambe was put in jail for two days. The charge: "obstructing development."

Construction of both the wind turbines and the Tarali Dam started after 1997. People displaced by the turbines are still awaiting the ever-elusive "rehabilitation." The company provided no jobs to the villagers, even to those whose lands had been acquired. And no one was ever compensated. Furthermore, the company does not supply any electricity to local consumers, and the project has so far not attracted any new opportunities or economic activities to the area. Rather than generating "greater local employment" as was claimed, it has in fact caused more unemployment and more misery in the area. The company maintains no contact with villagers. And while people are generally poor at Bhambe, after losing their lands to the project, the poverty has only deepened. As a result, villagers have said in unison that this wind project provided them with no benefits whatsoever.

Chikhali Village

At Chikhali, Ellora Times Ltd. took over 500 acres of villagers' lands at the ridiculous price of 16,000 rupees ($318 US) per acre. More than 150 wind turbines have since been raised at Chikhali.

Since the villagers of Chalkewadi (two kilometers from Chikhali) had supported MEDA in establishing the wind-energy plants, and Chalkewadi was much

Tata Motors, part of Tata Group, is the largest automobile manufacturer in India with revenues of $7.2 billion in 2007.

publicized as a model for such projects, Chikhali villagers thought they too would get good money and employment for a similar deal there. Villagers say a local trader acted as the agent for the company and convinced them to sell their "stony" land at near throwaway prices. The company even promised that the project would employ at least one member from each family that gave land away. According to the villagers, the company employed some locals in the beginning, but for a very short period, and now only two local people have permanent jobs with it; and they are security guards.

Chikhali has a small temple badly in need of repair. The villagers recall how, in a meeting held there, the company promised that all village children would become permanent employees of the firm. They also assured villagers they would build a new temple. But that was before the windmills. Now, if an animal strays onto grounds near the power plant, the guards will beat the creature. The company also did not pay the promised levy of 1,500 rupees ($30 US) per plant to the Gram Panchayat, and when they asked for it, the company literally told them to "go to hell."

Kadve Khurd

The project: a windmill turbine installed by Bharat Forge Ltd. (BFL), the flagship

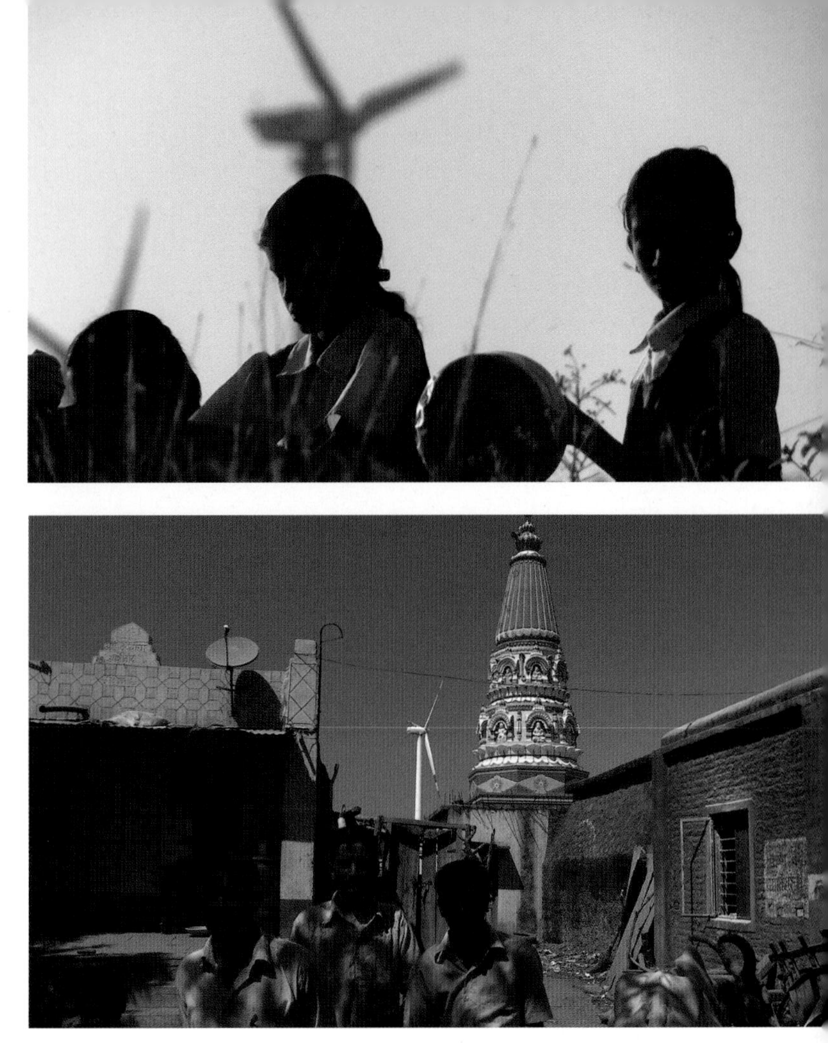

Windmills form the skyline in the region.

company of the $1.25-billion Kalyani Group, a "full service supplier" of engine and chassis components.

Kadve Khurd—a landlocked, isolated village—is part of the Sahyadri Valley located deep inside the forested hills of the Western Ghats. Today, 30 wind turbines stand in and around the village. The villagers knew nothing about the project before the Bharat Forge Ltd. started erecting turbines on their lands. To be sure, there was strong resistance by people whose lands were effected but, with support from the local administration that acted as an agent of the company, the plant was successfully constructed in 2001.

The project occupies 299 acres of land in the village, largely devottar (temple) properties, and some private farmland. This land deal was struck with a simple village headman whose family has been traditionally holding the land on behalf of the villagers. Because the villagers had no "official" titles to the land (they had old documents dating back to the 19th century and earlier), the company did not bother paying compensation. The local administration refused to hear the villagers' case, and in vain, they sought justice from the district collector's court in Pune. The district collector refused to place a stay order on construction of the plant and, in fact, annulled the stay given by a lower court. On top of this, the company, with support from police, slapped several agitating villagers with false charges of robbery and theft of equipment. It is no doubt that people in the village view the wind turbines as harmful junk which provide no benefit to anybody. It has given them neither electricity nor employment, and destroyed the only pasture in the village because the company has also put a total ban on cattle grazing in the project area.

The Story of Shivram Ahare

The company offered Shivram Ahare 50,000 rupees ($995 US) for his land, for which he had an old Sanad (grant deed). Shivram was not impressed. When all attempts at coaxing and bribing failed, the company threatened to kill him. He had to stay out of the village Kadve Khurd for two months. Shivram then met Vikram Shing Patankar, then minister of the Public Works Department in the Maharashtra state government. The minister had to ask the company to let Shivram alone.

Shivram was forced to file his first case against construction of the windmills in 2001 in the Tehsil (revenue block) Court, which declared Shivram's old land documents outdated. The case then went to the Sub-Divisional Court which stayed construction of the plants. The stay, as we saw, was later annulled through the district collector's intervention, and villagers allege that the district collector was heavily bribed. This belief is based on the fact that Shivram had an impressive list of original documents supporting his land claim, including a Sanad from the British period, along with an agricultural tax receipt, and original village land documents. All said, Shivram Ahare could not take his case to the High Court within the stipulated time frame because, by that time, all village records had been burned by company agents.

Supa and Satara

There is a project by Tata Motors Ltd. that involves grid-connected, wind-based electricity-generation facilities with aggregate capacity of 20.85 MW, located at two locations—Supa and Satara.

We visited the Supa windmill project at Sahajanpur, a remote and hilly village about eight kilometers from Supa. The village population is approximately 1,100, which represents some 200 families. The Project Design Document (given to the CDM authority) had previously announced it loud and clear that the company had no problem acquiring land from the villagers for the project, suggesting the people had willingly handed over the land. We talked to the villagers who told us that the windmill company had hired a few local residents to act as agents to facilitate the land acquisition. The agents would prepare the documents of sale and pay the villagers only 20,000 rupees ($400 US) per acre, which was much below the existing market rate for such fertile land. At the time we were there, the company had already acquired about 900 acres of village land from about 80 percent of the residents. Gulab Toloba Maiske, 65, said he was paid 60,000 rupees in 2001 for his three acres. The company officials told him they were in fact overpaying people in Sahajanpur by 8,000 rupees an acre since people in Satara had received only 12,000 rupees.

In Sahajanpur, the large landless scheduled-caste community was surviving on patches of government land, and all this, about 78 acres of it, was also captured for the wind farm. The community's livelihood was destroyed.

Jobs promised by the company to local

people have not materialized, save for some four or five people getting only some work. Sanjay Baba Mote, a local youth earlier employed in the company's Satara farm, was first transferred to Sahajanpur and then dismissed from his job. Mote said that there were many more like him who had met a similar fate. The company explained that the project was now complete and there was no more work for villagers. After having lost their lands, the only hope the villagers had was jobs in the company. And now it looks like nothing.

The Sarpanch of Sahajanpur, Vishwanath Laxman Shinde, who earlier worked as an agent for the company, stressed that the windmill projects had brought no benefit to the region. He said that the company had promised many things when it came, such as a 770-meter road, a building for the Mangal Karyalay (community hall), a two-floor building and vehicles for the Gram Panchayat, employment to the villagers, employment-generation programs, water ponds, electricity to the temple, and so on. In league with some influential and corrupt individuals in the area, the company didn't keep a single promise. The Sarpanch also complained that despite having acquired so much land from Sahajanpur, the wind farms were not paying any tax to the Gram Panchayat. Earlier, they had promised to pay 56,000 rupees ($1,115 US) to the Gram Panchayat in taxes every year. The villagers also said that with power cuts lasting from 8 to 12 hours a day, this windmill project was an utter failure. The Sarpanch is contemplating going to court to demand justice.

Jamdi, Phinalipada, and Jamde

A 6.25 MW windmill was installed by Godrej Industries Ltd. in the Jamdi village of Sakri Taluka. Two other small 2.5 MW mills were installed by Godrej Agrovet and Gulmohar Food and Feeds Ltd. in the Sakri Taluka villages of Phanalipada and Jamde respectively.

People in these three villages are not aware of CDM project concepts and the idea of carbon trading. Instead, villagers are worried about their livelihoods. The windmill companies promised to look after the villagers' social and economic welfare after the purchase of their land at rock-bottom prices. But the promises fail, welfare projects like hospitals are not erected, farmland is encroached upon, and grazing pastures are restricted. The poor become even more marginalized and in some of these areas villagers suffer from noise pollution and water depletions. All the while the companies that run these green CDM projects in Maharashtra State make big profits in the name of carbon emission reduction.

The Story of the Widow Farmer

We met 36-year-old Priti Vittal Pimapare, a widow farmer who belongs to a marginalized caste community in the village of Chaddvel. Her husband committed suicide in 2010 after harassment by representatives of a windmill company that had set up shop near the village. Their land had been fertile and was well positioned on the main road, but Priti's husband was forced

Vishanath Laxman Shinde, Chief of Nagar district, explains, "When our people go to the Shankar temple on the hill, they are targeted. They are called thieves, implicated in false charges, and put behind bars. We had asked for a school, but none of our demands have been met. They have not helped us in any way. Our children have no future. Our children are without education. We ask the state government to take pity on us and give us back some land to farm, so we may become slightly comfortable. Otherwise we have no alternative but to commit suicide. We may have to take poison. By 'we,' I mean the entire village. We are doomed because of this project. We have not benefited at all. We have been robbed. Our future generations have been destroyed."

to sell it at a very cheap rate. The company then put up a manufacturing and maintenance unit on it. Priti has two sons and one girl. The company promised a job for her one son but he still has not received any employment. Alone, with growing children, no land, no work, she gathered strength and asked the company for her land back. Nothing.

This is life under the windmills of Maharashtra State.

Acknowledgements to the publishers of earlier stories published in this work

1. *The Indian Clean Development Mechanism: Subsidizing and Legitimizing Corporate Pollution—An Overview of CDM in India with Case Studies from Various Sectors* NFFPFW (National Forum of Forest People and Forest Workers), NESPON and DISHA (Society for Direct Initiative for Social and Health Action), Nov. 2011, Compiled and Edited by Soumitra Ghosh and Subrat Kumar Sahu

2. *Critical Currents,* Dag Hammarskjöld Foundation Occasional Paper Series No.7, Nov. 2009, Carbon Trading—How It Works and Why It Fails.

CARBON OFFSET PROJECTS IN PANAMA

An Ironic Twist of Fate

by Carmencita Tedman MacIntyre
National Coordinator for the Defense of Lands and Waters, Panama

There are presently plans in various stages to develop more than 160 hydroelectric projects in Panama, where there is absolutely no need for them.

The United Nation's promotion of carbon offset projects as a way to prevent or reduce global warming and save our environment is one great farce. We have seen it play out here, in Panama.

It is very disheartening that we, environmental activists, have had to take a major stake in the fight against projects which classify as Clean Development Mechanisms (CDMs) under the United Nations Framework Convention on Climate Change (UNFCCC). The CDM allows a country with an emission-reduction or emission-limitation commitment under the Kyoto Protocol to implement an emission-reduction project in developing countries like Panama. Ironically, the CDM has become in a very short period the greatest source of destruction of nature and pristine ecosystems, rivers, and animal habitats that we have ever experienced in this country.

CDM projects can earn saleable Certified Emission Reduction (CER) credits, each equivalent to one ton of CO_2. These credits can then be counted toward meeting Kyoto targets. Here is where a plan that seemed well intended in the beginning has backfired and gone completely wrong around the world. The situation in Panama is a prime example of it.

Actually, the consequences of the drive to implement CDM projects are so bad that Panama is now much more rapidly

contributing toward global warming, emitting more polluting greenhouse gases, significantly decreasing its green coverage, destroying its biodiversity, cutting down protected ancient forests, annihilating hydrographic basins, drying up wetlands, eliminating animal habitats, destroying mangrove forests, and negatively affecting both the Atlantic and the Pacific oceans. Sadly, all this damage is irreversible. In the process, indigenous peoples' human rights have been violated, peasants have been abused, and farmers' and citizens' civil rights have been negated by government corruption, police violence, and corporate greed. And all this has been propelled by the corporate carbon rush craze being pushed by the UN. What for? Not to save our planet—but for profit. And Panama has become another victim of it.

The Republic of Panama approved the ratification of the UNFCCC in 1995. Panama then approved the ratification of the Kyoto Protocol in 1998. In order to implement related project activities that result in recognized CERs of greenhouse gas (GHG), Panama's National Authority of the Environment (ANAM) has opened the doors to corporations, mostly powerful transnationals, to take hold of large parts of our small country and to develop projects, all supposedly to gain CDM registration. Currently, such projects announced in Panama are 95% hydroelectric, plus some eolic, biomass energy, solid waste management, alternative fuel, reforesting, and transportation schemes. The problem is that in accommodating these powerful international corporations, environmental and property laws have been completely relaxed. Previously protected forests have become vulnerable, and citizens have lost rights to their land. The ANAM has effectively become the authority to grant permits to destroy the environment, while taking zero measures to mitigate climate change or even address the adverse impacts of these CDM projects, none of which could be considered "sustainable development." The original theory behind carbon offsets was that GHG reductions would result in fewer emissions entering the atmosphere from industrialized countries. The more heavily polluting

developed countries would buy carbon offsets to fund "green" projects in cheaper developing nations, with an idea that it was easier to push projects in areas without developed infrastructure. However, the ease with which corporate-backed CDM projects trample over and disregard the constitutional laws and rights of citizens and continue to destroy the environment in developing countries makes it clear that this has been only pretense.

In Panama, every single hydroelectric project promoted falsely as a mechanism to develop clean energy has been a disaster.

Indigenous Land Rights Abuses and Territorial Conflicts Caused by Projects Promoted as CDMs

The indigenous peoples of Panama have all fought extensively for their rights, and now some groups have the recognized rights to collective territories in the form of *Comarcas*, identified legally by governments after more than 500 hundred years of oppression. Now, in post-Colonial days, the tyranny, hardship, and suffering continues in the hands of transnationals pushing CDM projects into indigenous territories and also into protected national and international parks which neighbor indigenous lands.

There are currently firms pushing into the Tabasará River, Changuinola River, Bonyic River in Naso Tjerdi Territory, Chorcha River, Fonseca River, and elsewhere that are affecting indigenous territories, yet seeking CDM registration.

The Naso People of the Naso Tjerdi indigenous territory have fought for years to defend their ancestral territory, to defend their Teribe River, the Bonyic River, and all their waterways and forests against land encroachment by cattle ranchers

and, most of all, to protest against the Bonyic hydroelectric project. The Naso have fought a dignified battle against the half-built hydroelectric project. They implemented watch camps on the river, set up camps that lasted months outside the Presidential Palace in Panama City, denounced Panama for human rights violations at international human rights courts, did all humanly possible to stop the project. The Panamanian government and the corporation behind the project instead persecuted the Naso, set up legal battles against Naso leaders through judicial courts and false charges, jailed them, employed police brutality, tear gas bombs, and shot guns. And the project advanced with armed workers who invaded Naso territory.

Naso ancestral lands are located in the heart of the La Amistad Biosphere Reserve and form part of the International Park "La Amistad," a World Heritage Site, and the Bosque Protector Palo Seco National Forest, comprising a paradise of biodiversity, with many endangered species and endemic plant and animal life not found anywhere else on Earth. The Naso and their natural habitat form an extremely important part of the Mesoamerican Biological Corridor. The Naso are themselves an endangered human species. There are only about 3,000 Naso left on the planet, not a sustainable population in the face of the pressures they suffer today. The Naso are beautiful, strong, and kind with a wonderful culture of respect and admiration for nature. They have a language, traditions, and a culture that we must save, because their presence in this world makes it better. But the Naso face extinction as a people if the hydroelectric projects continue and their land tenure rights are not respected.

The Bonyic hydroelectric project on such pristine ecosystems is heartbreaking

ARE

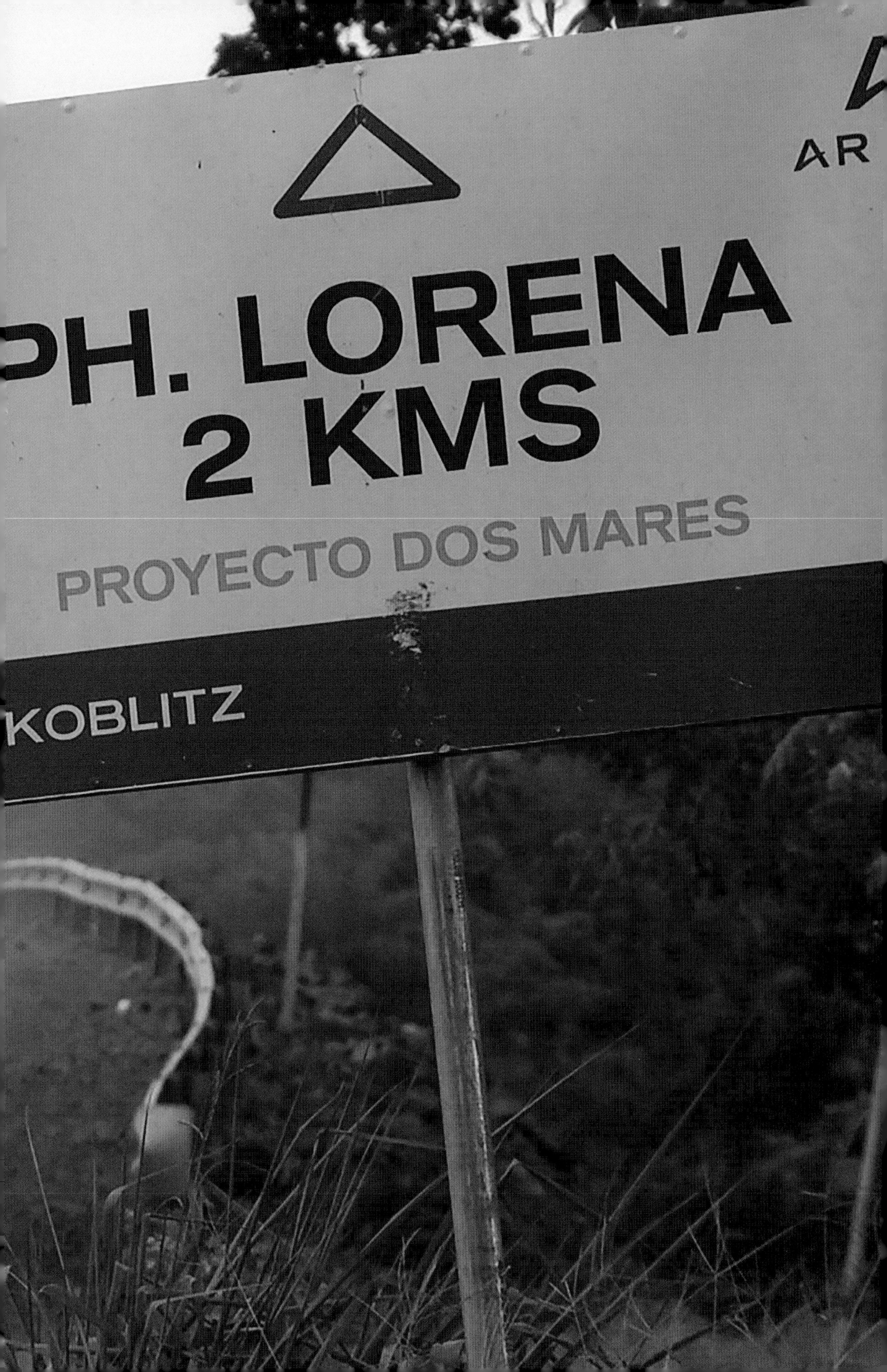

PH. LORENA
2 KMS
PROYECTO DOS MARES
KOBLITZ

and should have never been conceived. Nevertheless, it is currently at the validation stage for CDM registration. This project is a crime against humanity. It should be stopped and dismantled.

The Ngöbe-Buglé people of the ÑoKribo region fought hard against invasion of their lands by the Changuinola River Basin hydroelectric projects promoted by AES Corporation, a huge transnational based in the US. Despite all efforts against it, the Changuinola I (Chan 75) dam was completed in mid-2011. Ironically, these projects of massive destruction are located in the International Biosphere Reserve La Amistad. La Amistad was declared a Patrimony of Humanity and World Heritage site by UNESCO in 1990. Also a protected forest called "Bosque Protector Palo Seco" inside the Biosphere Reserve has been damaged by these projects. There were great protests against construction of Changuinola I, including camps set up in Panama City, suits at the International Human Rights Courts, a long march lasting twenty-two days from Bocas del Toro, Chiriquí, and the Comarca Ngöbe-Buglé to Panama City. But governments did not listen. Instead, the Panamanian government violently crushed protesters, humiliated and jailed indigenous men, women, and children, and forcefully evicted the Ngöbe from their homes with tear gas bombs and lead bullets. A sweeping persecution was mounted. Dissidents were threatened. The zone was militarized and the project proceeded under the steely gaze of contracted foreign engineers, who supervised massive devastation of the forest, destruction of the once great Changuinola River, and the ravage of endemic and native animal habitats. A few months ago, upon finishing construction of the dam, thousands of fish and aquatic species died. ANAM did absolutely nothing despite the fact that AES did not implement a responsible mitigation

"They have privatized the water system which forbids people from fishing, from drinking the water so that they will all be able to get carbon credits."

plan to save the species. Still, the fight continues against the other two dam projects on the same river.

The Ngöbe-Buglé people of the Kodrini Region have fought against the Barro Blanco dam project across the Tabasará River for 12 years. Sadly, the area was militarized and heavy equipment forcefully entered the Tabasará River region. The associated Environmental Impact Assessment (EIA) study lacked veracity and purposefully left out the fact that the project would affect indigenous peoples. This was done to better attract international funding. There was never any public consultation with indigenous communities who were never invited to attend the Public Forum. Now armed national police and armed guards defend the project against native peoples who are trying to protect their rights. But this is not new to the Ngöbe-Buglé. It is well known here that the original Tabasará dam was planned many years ago to supply energy to the highly controversial Cerro Colorado Copper Mine inside Ngöbe territory, where lies the second largest unexploited copper deposit in the world.

The Chorcha hydroelectric project across the Chorcha River has been promoted by Aguas y Energia, a Colombian corporation. The Ngöbe-Buglé of the Nedrini region have been fighting against it since 2006. At the Public Forum in November 2007, more than 1,000 people voiced unanimous opposition. Nevertheless, ANAM and the government gave approval to the project despite the ill-constructed EIA study. On December 1st, 2009, in Plan de Chorcha, all rubber boots donated by the company were gathered and set on fire to protest. The Immediate Chief Mercedes Mendoza, a strong woman, said, "A long time ago my people were sold out for mirrors. Today I and my people will not be bought for rubber boots." That same day a walking expedition set out toward Quebrada Arenas, the site of the proposed hydroelectric project. After hours of walking, the last group of about 17 people, com-

prised of women, children, and the elderly, accompanied by the Cacique Celestino Mariano Gallardo, and the Immediate Chief Mercedes Mendoza were intercepted in the dark of night by riot police, who lined up in formation to attack. The captain shouted at the unsuspecting walkers that he had orders to shoot at "anything that moved." To that we answered, "These are ancient indigenous footpaths used since ancestral times. Here are the original Panamanian people, the first people. You, police captain, are in a gray area without jurisdiction, the Comarca markers have been removed and our limits violated, yet you protect a foreign company against the original people." Women and children have since set up camp on the river bank to watch over it. The fight continues to defend the Chorcha River, and the mountain range against development of the Chorcha Guarivira Open Pit Mine.

The Government Starts a War against the Indigenous People

On January 30th 2012, the Ngöbe and Buglé people led a countrywide peaceful protest against mines and hydroelectric projects in the Comarca and indigenous territories, blockading the Inter-American Highway and other inter-provincial roads, bringing all traffic to a complete stop for six days, and at several points, leading to the greatest economic disruption of the country since the military dictatorship. This action gained the moral support of most of the Panamanian people.

On February 5th, national police forces, specialized frontier units, and riot police

fiercefully attacked the defenseless protesters by air, from helicopters, and by land.

On that day the war for water and the environment officially started in Panama. The protestors, in great part women, children, and youth, were indiscriminately massacred and attacked with tear gas, chemicals, brutal and lethal force by police at every site of the protest, leaving three Ngöbe dead. My friend, 26-year-old Ngöbe Jeronimo Rodriguez Tugri, and a 16-year-old mentally challenged boy, Mauricio Mendez, were assassinated by the police. Detained Ngöbe women and teenagers were gang raped and molested by police units. Ngöbe children were jailed and hundreds of people imprisoned without charge. Dozens of people were severely maimed and injured, with many losing body parts and eyesight. The Martinelli government and police violated all kinds of human rights and committed great crimes against humanity.

On the 8th of February, a government-Ngöbe dialogue started. There were all kinds of manipulations and betrayals. The Catholic Church and the United Nations (UN) became involved in the talks, supposedly to ensure proper mediation. However, they had conflicts of interest, especially the UN, which had given carbon credits to the controversial Barro Blanco Hydroelectric Project over the Tabasará River, impacting Ngöbe communities. When the final accord was signed it was a great deception for most of the Ngöbe people. The President of the traditional Ngöbe-Buglé and Campesino Congress,

Celio Guerra, who protested alongside his people, was never allowed a voting voice in the dialogue. The big scandal was that without consulting the true Ngöbe protest leaders and bases, some self-imposed Ngöbe negotiators and their assessors signed an accord behind the protestors' back. This would allow future hydroelectric projects in the Comarca. All these underhanded dealings have infuriated the betrayed Ngöbe people who are currently re-organizing to continue the resistance and fight against these projects.

Injustice Continues after Projects Are Completed

The Esti hydroelectric project, which began in 2003 utilizing the flow of five rivers to produce electricity, dumps all waters in cycles into the Esti River bed which has insufficient capacity to handle the volume. The consequence has been flash floods and the destruction of kilometers of gallery forests along the Chiriquí and Esti rivers. Other rivers and creeks that were deviated for the project are drying up. Farmers, like Mr. Agripino Carrillo and his family, have battled for years with law suits in an ongoing fight against promoters of these projects that illegally cut fences and enter private property, destroy rivers, cause the collapse of very expensive cattle breeding projects, and trigger flash floods, landslides, and the subsequent loss of hectares of agricultural soil, not to mention drowned cattle and horses. At times farms are split leaving areas without communication, wells dried up, and human life endangered. In this battle, only a few economically capable farmers were able to hire lawyers to defend their rights. Even then there are countless testimonies of corruption and underhanded dealings, where either lawyers sell out or court magistrates unjustly and against reason stand on the side of the transnationals until most people just give up.

Peasant farmers without the ability to hire anyone to defend their rights were forced to give up their land and livelihoods, and they were inadequately "compensated" with a misery of pennies per square meter. To add insult to injury, after years of misappropriating extensive pieces of land way more than they needed for the project, these corporations are now into the real estate business, promoting beautiful properties for sale around the Barrigon artificial lake at top dollar to North American and European retirees, while the people that used to own the land only look into the distance, completely stripped of what they held most dear. On top of it all, the people are continuously confronted by ever-present signs that read, "PRIVATE PROPERTY", "YOU ARE BEING WATCHED BY CAMERAS", "NO SWIMMING", "NO FISHING." They must also cross through armed gate-points on the public bridge and road. All this is effectively the privatization of rivers, the privatization of water.

Then, to make matters even worse, one company, as part of its mitigation plan, has reforested with teak, an aggressive Asian tree species, that does not allow native species to grow under it. Native animals cannot feed from it. However, much money can be made from the monoculture of this invasive species—masked as reforestation—because the highly sought-after wood can be cut and sold once the trees mature. This cannot be called reforestation. It's dirty energy.

During more than a year, the Esti project has been re-building collapsed tunnels and infrastructure. The communities are complaining and organizing against the transnational because for many months

it has been dumping into the river water that is contaminated with construction waste and chemicals. This happens a short distance up from the community´s drinking water intake.

A Case Study of How Hydroelectric Projects Proceed through Construction—Dos Mares

The Dos Mares hydroelectric project, still under construction in 2012, plans to take up the Esti project waters through a system of several dams and power canals that guide the waters through two districts comprising three hydroelectrics: the Gualaca project (19.8 MW), the Lorena project (33.8 MW), and the Prudencia project (58.5 MW). The promoters for Dos Mares project are European transnationals, and the major financing institution is the European Investment Bank (EIB), using European Union public money. The main contractor is a Brazilian firm (Odebrecht) that has faced a myriad of accusations and law suits in countries where it has worked. Such Not-So-Clean Mechanisms as this are truly global enterprises that attract big international money.

Like the other hydroelectric projects in Panama, Dos Mares has been plagued by a spectrum of accusations including a lack of transparency, abuse of power, harassment and coercion of farmers, invasion and seizure of private property, forced expropriations of land, abuse of workers' rights, endangering and destroying people's possessions, supplies and livestock, disrupting people's lives, contaminating water systems, endangering the health of communities, destroying valuable ecosystems, devastating wetlands, eliminating animal habitats, ending aquatic animal life-cycles, and killing mangrove systems at the Pacific Ocean.

Nevertheless, Dos Mares is at the CDM validation process stage with the UNFCCC. The problem with this process is that the people most impacted by these projects have no effective voice in the dealings because few, if any, even know about it. They are in English, a foreign language,

and local people simply do not have access to any of the information. The local population is further disappointed that promises made by Dos Mares to help and work for the communities have not been fulfilled. The project is near its completion and, because of all the abuse, suffering, and disruption to communities and environment, people here feel they have gotten a raw deal. This unfair treatment is from a project with a total cost of $422 million US.

Unfair Land Acquisitions

The Dos Mares issue over unfair land acquisitions is not over. Farmers testify that Dos Mares representatives visited them separately and individually and threatened that if they did not surrender their lands, and accepted their arbitrary, well-below-market-value offer, Dos Mares would take them into an expropriation process through the ASEP (Authority of Public Services), which is the governmental entity that gives concessions for hydroelectric projects. They were told that if such an expropriation process went through, they would get nothing at all for their land. Dos Mares took advantage of people's ignorance of the law to intimidate and coerce them into signing unfair contracts out of fear. Throughout all these unjust practices the Panamanian government turned a blind eye, and in some cases, even aided the process.

All of these threats were based on a deliberate misinterpretation of Panamanian law. The law actually says that if the government declares a scheme a "Social Interest Project," the easements could be forced and the lands expropriated *if* there is no other way to advance the project. Nevertheless, the law states that in the event of expropriation, owners of the land must be fairly compensated at current market value or the value of the loss of business. Indeed, if the people had become orga-

nized and fought Dos Mares, everyone would have ended up getting a much better compensation package for their properties than the ridiculous amounts most people received. In fact, many farmers are still in limbo because Dos Mares invaded their properties without going through the proper legal framework, destroyed the lands, and have not paid them anything at all. Only a few others are awaiting trials and outcomes of legal and administrative proceedings through the Panamanian Judicial Courts and ASEP. In one case the ASEP cited and agreed to the price requested by the farmer as fair, but Dos Mares has refused to pay. There is obviously no true justice in Panama.

Destruction of Natural and Aquatic Habitats

The case of the Ortega family is particularly shocking. Beltrana Ortega, a retired high school principal, took it upon herself to find some type of justice and assert her family's civil and property rights under the Panamanian Constitution. Beltrana engaged in a series of meetings and correspondences that went back and forth between Dos Mares, Odebrecht (the Brazilian contracting firm) and ASEP. She argued that the practices employed through the project were unfair and she was trying to get some justice. The Ortegas are a humble hard-working family that owns probably the most valuably located property within the entire project plan. It is right alongside the Inter-American Highway, minutes from David, the capital of the province, seconds from the Chiriquí River, and alongside a creek and near the mangrove forests. For 50 years Beltrana's father had protected these wetlands, an untouched habitat for many land and aquatic species like alligator, turtle, duck, guichichi, green iguana (endangered species), brown iguana, and many species of bird, frog, snake, butterfly and insect.

Also, it was the only natural nursery in the area for many species of diadromous fish, and other aquatic species that need to migrate over their life cycle from mangrove to fresh waters. These include the robalo, sabalo, chupapiedra, crab, and fresh water giant shrimp which are becoming extinct.

One day the Ortegas were surprised when suddenly and without prior notice, Odebrecht, under Dos Mares directives, invaded the Ortega's protected wetlands with heavy equipment, turned everything upside down and made a muddy mess. In tears and astonishment they say how project workers scooped up the dying fish, shrimp, and other species by the thousands in buckets. Meanwhile the area was heavily guarded by Panamanian National Police and armed Dos Mares guards preventing any civilians from approaching and documenting what was happening. Beltrana never gets over the shock of the impotence she felt and has tears in her eyes every time she recalls this story. But this is only one of many such stories right here in Panama.

Dos Mares committed great crimes against animal life and habitats all along the project, drying up waterways, breaking up migratory passage ways. Dos Mares workers tell of an incident where heavy equipment carelessly was sent into animal habitat and broke a big alligator's back, after which the boss ordered them to make a big hole with the backhoe and bury it alive.

Supposedly Dos Mares had to engage in a reforestation program, but no one has seen them reforest one single tree. What they did was donate some trees to the Elementary School of Higueron, took some pictures, and left the children and teachers to plant these trees without fertilizers or follow-up during the dry season, and as a consequence when the new school year started, 97% of the trees had died. Dos Mares never came back to check.

It is important to note that the lending institutions, like the European Investment Bank (EIB), are equally to blame for all these terrible wrongdoings because they do not responsibly follow up to verify that the projects are complying with the banks' established regulations.

Attempts against Human Life, Communities, Public Safety, Private Property, Domestic Animals, and Farms

During construction of the first phase of the Gualaca hydroelectric project, a kilometers-long power canal was built that led to the first engine house. One day the regular rains of the season came and the canal quickly filled to levels where there was danger that it would spill over to the construction site. So the contractor opened an enormous breach in the canal. A sea of water spilled into a tiny creek and over pasture lands and into the community of Higueron. Farmers suddenly found their homes and stables flooded, with the water drowning pigs, chickens, and swamping their feed storage sheds. There was never any prior warning that workers were planning to flood the community to save the company's investments. The people were forced to take their children and horses and run for their lives with only their clothes on. The elementary school was flooded and the school bus could not pass through the flood waters, so the children had to take off their shoes and risk their lives crossing the flood waters. The humble people of Higueron lost a lot: their domestic animals, yucca crops (main starch food), beans, corn, construction materials, mattresses, chairs, furniture, stoves, fridges, washing ma-

chines. Everything that got wet was damaged, including their homes which were filled to knee height with mud. A horrified people organized a protest and demanded compensation from the transnational. The people took to the project gates and picketed and protested. The governor of the province was called along with the mayor of Gualaca and other municipal representatives. Odebrecht's representative said they were not responsible for damages because they were only the contractor for Dos Mares. The Dos Mares representative said that Odebrecht was the one that had opened the hole. Then, after pressure from the community and in front of the governmental authorities, the Dos Mares representative said they were aware of all the damage done but they did not have anything in their budget to pay compensation for such an unforeseen event. And that was the end of that. The Panamanian authorities did not seek justice for its people, the community went back to their homes poorer than they were, and had to work hard in efforts to replace what they could of their losses. How could the UN promote these projects as "sustainable development" and clean energy?

Sickness, Disease, and Social and Economic Disturbance from Hydroelectric Construction

In the meantime, for months on end, 24 hours a day, 7 days a week, babies, children, elderly, women, men, and animals were living in horrible conditions surrounded by a cloud of dust, drumming, thumping and other loud noises, plus aggressive driving of project trucks on public roads. Children and adults complained of skin, ear, eye, lung, and respiratory infections and diseases from the high amount of dust they were breathing. Also, the situation created a very high level of tension, stress, anxiety, and panic attacks among people living near the project. An elderly man and his wife had to be treated by a psychiatrist for depression and traumatic stress syndrome. Eventually, the elderly couple was forced to abandon their home for many months as they could not withstand the related stress and disease. This couple initiated a legal suit against the project for damages, but their case was not answered by the Panamanian courts and they have had no justice.

Cattle grazing and feeding were also disturbed by the dust and noise which affected the quality of milk and animal growth patterns. Also, the canals split the entire territory, forcing farmers to take lengthy detours to get to their land and animals. Dos Mares had promised a proper bridge to handle vehicles and animal crossings, but only delivered a hanging one along a footpath.

Furthermore, dozens of homes and buildings were damaged by cracks on foundations, floors, walls, and supportive beams due to explosive detonations and earth moving. The company refused to pay for damages.

Public Consultations Failure

In order for a project to be approved for CDM registration, it must undergo public consultation with those who would be directly affected by it. Dos Mares organized unannounced public forums for Gualaca, Lorena, and Prudencia. Hence, many people who would be directly affected never attended. The ones who made it to the meetings complained they were manipulated, with the company not answering questions from the people. In fact, they were not allowed to express their opinions, or ask questions out loud. They were instead instructed to write their queries on pieces of paper that were then selected for being answered by corporate representatives if they found them agreeable.

Hydroelectric Madness in Panama

There are presently plans in various stages to develop more than 160 hydroelectric projects in this tiny country. All projects are planned on pristine, crystalline rivers in tropical humid forests and virtually untouched habitats, all of which will be irreversibly destroyed. There have never been studies of the cumulative and synergic impacts of all these projects. As such there have been excesses at places like the Chiriquí Viejo River basin which has 26 projects on it.

In Panama, there is absolutely no need for this: even by 2006 the country already had the installed capacity to produce 1,925 MW of electricity, and in the year 2012 peak national consumption reached top levels of 1,125 MW. So Panama now produces an excess of electricity which is privately sold to Central America more cheaply than it is sold at home. Every day private corporations conduct net exports of electricity from Panama, through the Central American System of Electrical Interconnection (SIEPAC), which travels through a network of high-power lines that have in themselves become sources of much grief and destruction.

Obviously, *none* of these proposed electrical projects are needed in Panama.

***

The failure of the carbon offset system is more than apparent here. In Panama, CDM projects are increasing the emissions of carbon dioxide and greenhouse gases into the atmosphere, and contributing to climate change. This is because there is a rush to develop infrastructure and projects that are otherwise completely unneeded. These projects develop at the expense of pristine tropical ecosystems. And bafflingly, emissions elsewhere in the world are not reduced since organizations in industrialized countries can buy CERs in developing countries such as Panama. It's truly an ironic twist of fate.

PICKING THE WRONG TEAM

Land Conflict in Honduras's Aguán Valley

by Jesse Freeston
Journalist and Film Maker

Dinant-Exportadora's biogas recovery plant using African palm oil was approved for carbon credits in July 2011 despite the fact that numerous human rights organizations have found the owner of Dinant Corporation Miguel Facussé with his hired private security force to be implicated in the assassination of at least 31 local farmers since November 2010.

The land conflict going on in Honduras's Aguán Valley, while as complex as any other conflict, can be boiled down to two camps representing drastically different visions for the Honduran countryside. One group, headed by a handful of business magnates, sees the land as key to Honduras's future as a world leader in palm oil production, for use as biofuel and vegetable oil. This group, known as the *empresarios,* see their profits as an appropriate reward for the wisdom and courage they've shown in investing in palm oil. The other group, headed by thousands of organized farming families, sees the land as a provider of a dignified life for everyone in the valley. This group of small-scale and landless farmers, known as *campesinos,* have proven they are willing to die for their vision, and the *empresarios* have shown that they are willing to kill for theirs. In July of 2011, for instance, the UN's Clean Development Mechanism (CDM) approved carbon credits for an *empresario* alleged to have killed more than 40 *campesinos* over two years.

Palm oil is the world's preferred vegetable oil, representing one third of the global market. Its growth in popularity over the last few decades is considered by many to be the fastest expansion of a single crop in world history. Today it can be found in ice cream, margarine, cosmetics, soaps, and much much more. A 2009 study by the

The red fruit of the palm is loaded for transport.

Francisco Ramírez, a peasant who has been occupying land belonging to Miguel Facussé, explains the violence he has lived through. "The problem we had just recently was when we arrived to work on those lands that belong to Miguel Facussé we found hitmen paid by him, lying on the ground, and that's when they attacked us with gunfire. They began to shoot at us and five comrades were killed, and others were wounded. I myself was wounded in that incident and as proof I have a wound where a bullet hit my face on the right side and went out on the left, taking all my teeth with it. These are the consequences that we suffer in this country simply because of the need for a piece of land."

UK newspaper, *The Independent*, found that 43 of Britain's top 100 grocery brands contain palm oil. Forecasts now indicate that global demand for the thick syrup-like stuff will continue to surge as investors further look to palm oil as a fuel source to supplement petroleum.

In 2009, the sale of palm oil topped $20 billion globally, and for the last 20 years, Honduras's richest person, Miguel Facussé, has been increasing his share of that pie. Facussé's palm plantations cover roughly one-third of the arable land in the Aguán Valley. His guards carry automatic weapons, drive unmarked pick up trucks with no plates, and typically wear ski-masks to hide their faces. They have been implicated in the killings of at least 40 campesinos in the Aguán in the last two years alone, campesinos that have organized to challenge Facussé for control of the land.

The Butcher of the Aguán

On November 15th, 2010, five campesinos, members of Aguán Campesino Movement (MCA by its Spanish initials), were shot dead on their way to work. On the day after the massacre, Facussé appeared on national television admitting that his guards killed them, but that they were defending themselves from an ambush by hundreds of campesinos with AK-47s. Both the police and the district attorney said they believed Facussé's story, and offered a photo of a lone AK-47 found at the scene. However, they did admit that Facussé's guards controlled that crime scene for over an hour afterwards until authorities finally arrived.

MCA has been living and working on these lands for almost 12 years. The area used to belong to the Regional Military Training Center (CREM by its Spanish initials), a military base set up by the Pentagon in the 1980s and used primarily to arm and train the counter-revolutionary army in neighboring Nicaragua. But the military base was also active in "counter-insurgency" activities in the Aguán as well. "They used to torture our comrades right here," recounts MCA member Adolfo Cruz, "but today we practice life where they once practiced death." All that remains of the base itself are the concrete foundations, but the US-trained soldiers, they're still here.

The Honduran military maintains a small encampment directly across from the elementary school inside the MCA community of Guadalupe Carney. A few hundred feet from the field where the campesino children play soccer and tag, live soldiers collecting intelligence on the campesinos and, according to Cruz, trying to scare the farmers into submission. The government says the troops are there to find the rest of the campesinos' AK-47s. But after more than a year, they've yet to find a single weapon. This kind of militarization has become the norm in the Aguán, ever since the President of Honduras was kidnapped in his pajamas by the army on June 28th, 2009.

The Coup

At 5 a.m., Honduran President Manuel Zelaya was taken out of his house at gunpoint by soldiers, flown to Costa Rica, and left on the airport runway. The date was significant; it was to be the first time in Honduran history that Hondurans were to vote on something other than who would govern them: Zelaya was holding a non-binding survey on whether Hondurans wanted to rewrite the constitution.

With Zelaya safely out of the country, members of congress produced a forged resignation letter and voted congressman Roberto Micheletti the new president. Meanwhile, the military was deployed

throughout the country to take control of the streets. Those who reported on the coup saw their television and radio stations taken over by the military. Dissenting journalists were taken into detention. All the while the corporate media pretended nothing was happening and aired soap operas and cartoons.

Elected in 2005 as a member of the pro-business Liberal Party, few in Honduras expected that Zelaya, himself the son of a wealthy family dating back to Spanish colonization, was going to depart from Honduran politics-as-usual. But by his second year in office, Zelaya had turned the Presidential Palace—long considered closed to the public—into a place where Hondurans were encouraged to come and discuss their problems and proposals with the President's staff. By his third year he had dramatically raised the minimum wage, supported community-run hospitals for Honduras's long-marginalized black communities, put a moratorium on unpopular mining concessions, and joined ALBA, a trade pact among left-leaning Latin American governments that stresses mutual aid, barter trade, and poverty eradication, an alternative to the West-led "free-trade agreements." In 2008, Zelaya passed a Presidential decree declaring that any campesino group that had maintained a productive occupation for 10 years or more would be given the opportunity to buy titles to the land, mortgaged at no interest by the government. MCA was one of the first in line for this new opportunity.

The campesinos of the Aguán also joined in Zelaya's most far-reaching proposal for a new, more participatory constitution. The political system generated by the outdated constitution, written under US-backed military dictatorship in the early 1980s, had proven itself incapable of responding to the demands of Honduras's poor majority. Over 30 years of electoral

On June 28th, 2009, the National People's Resistance Front was born. Initially a movement against the Honduran coup d'état, the organization worked for the return of President Jose Manuel Zelaya Rosales and democracy. The resistance movement was always broad based and included fighting for the return of land for the landless peasants in the Aguán region.

democracy, Honduran society has become more violent and more unequal. But the people were stuck in a bind because the constitution made participation in referendums and other kinds of direct democracy illegal. So Zelaya proposed creation of a Constitutional Assembly that would include representation from all sectors of Honduran society. Together they would write a new constitution. They would "re-found" Honduras. However, when he announced a referendum on this new Constitutional Assembly, the Supreme Court said such activity would be unconstitutional. When he organized a non-binding survey on the matter, the country's elites and the military united to overthrow him.

The FNRP is Born

The subsequent military coup gave birth to a new political force in the country, the National People's Resistance Front (FNRP by its Spanish initials). The FNRP is the organizational structure that developed out of the spontaneous popular uprising against the coup. Today, it has chapters in all Honduras's 298 municipalities, including organizations that represent virtually every sector of Honduran society. Teachers, street vendors, taxi drivers, sex workers, students, LGBT rights defenders, feminists, industrial workers, indigenous rights groups, black unity organizations, lawyers, small- and medium-sized business owners, nurses, and of course campesinos—these are just some of the groups you will find represented at FNRP meetings.

The campesinos of the Aguán have a special status within the FNRP. They are highly respected and generate both inspiration and sympathy, for the success of their land occupations and the repression that they've faced. The first-ever national assembly of the FNRP was called the "Martyrs of the Aguán National Assembly" in recognition of their sacrifices.

Orvelina Flores Hernades, a member of the Aguán Campesinos Movement who has been occupying land, explains her point of view regarding who is receiving money from carbon trading, "On my way of thinking, it's an injustice that they are receiving money because maybe they are not using that money for a good cause, but rather to take the lives of poor peasants. We have the right to a better life as Hondurans, as people and it's an injustice on the part of foreign countries, to give money to that man, instead to us, the poor, who need that money."

Investors are now looking to palm oil biofuel to support growing demand for fossil fuels. Pressure on lands in the Global South will only increase as a result.

In the months following the coup, the FNRP had two central demands. The first was the return of Zelaya to power. The second was for a referendum on the Constitutional Assembly (this was supposed to have taken place on the day the coup occurred). Instead, Honduras got a presidential election run by the same military that overthrew President Zelaya five months earlier. And the only candidate "running" at the end of it all was a pro-coup supporter.

Coup under Construction

In the months leading up to the election, more than 4,000 Hondurans were jailed without charges, dozens of activists were assassinated, protesters were routinely attacked by police and military, anti-coup media outlets were shut down, and the only anti-coup presidential candidate had his arm broken while being arrested by police. He later renounced his candidacy, calling the elections a farce. Three weeks before the election, the military sent a letter to all mayors in the country demanding that they hand over the contact information for all known FNRP members in their jurisdiction. At the same time, the military called in the reserves, an action that the constitution only permits during times of war. One week before the election, the generals made a series of statements to the media that anyone opposing the elections would

Pruning palm trees in the monoculture forest.

Wilfred Paz, Coordinator for the Aguán Valley Permanent Human Rights Observatory, explains the role of shifting community markets in fueling the land conflict in Honduras, " We saw a global oil crisis begin and biodiesel production appears as an alternative. Through the initiative of biodiesel production, the lands of the Aguán acquire a geo-strategic, political and economical status because the businesspeople of the lands see that by planting palm trees in sufficient quantities, they can become the future oil producers of the nation and of Latin America."

Carbon emissions bellow from the processing plants, from stacks, and from cookers, at night.

be tried for treason and face up to 70 years in prison.

The Organization of American States, the Carter Center, the EU, and UN, the world's four major election monitoring bodies, all refused to observe the elections. Both the FNRP inside Honduras and countless observers outside screamed that the elections were nothing but coup-laundering. Nevertheless, Hilary Clinton's State Department announced its support, providing "observers" through its subsidiaries: the International Republican Institute and the National Democratic Institute. The *New York Times*, despite not having a single journalist based in Central America, wrote an unsigned editorial declaring "wide agreement" that the election "was clean and fair." The upshot: the ultra-right Porfirio "Pepe" Lobo, who had lost to Zelaya in the 2005 election, was lauded as the new democratic leader of Honduras. Canada, Costa Rica, Panama, Peru, Colombia and, most importantly by far, the United States recognized the elections as legitimate, and restored normal relations with the Honduran government. The world looked away. The coup-plotters, known across Latin America as *golpistas*, had won another chapter of what the FNRP calls a "coup under construction."

With their hopes for peacefully restoring Zelaya to power now dashed, the previously unknown Unified Campesino Movement of the Aguán (MUCA by its Spanish initials) launched the largest single land occupation in the country's history. Just ten days after the election, they took over 5,000 hectares of Facussé's palm plantations.

Meanwhile, the Lobo administration quickly went to work undoing progressive laws passed under Zelaya. They also began privatizing the education system, increasing funding to the military and police, they changed the constitution to allow military to do police work, signed a free-trade agreement with Canada, passed an anti-terrorism law, made the morning-after pill illegal, and launched "Guardians of the Homeland," a military

Community Organizer Mario Mumbreno draws the links between the production of palm oil and carbon credits, "There is a kind of collective consciousness here around the fact that the palm corporations are taking advantage of a certain control, or a political and economic influence they have at the level of central government, in order to end up with these benefits, which are in exchange for being mindful of carbon. The famous carbon credits. Generally, the beneficiaries of this system are not going to be the communities."

training program for children from poor communities. As for the campesinos, Zelaya's minor land reform decree was declared unconstitutional. MCA and dozens of other communities never got their land titles. When they blockaded highways in protest, Lobo sent in the military.

The civilian and military leaders of the coup have been rewarded under Lobo. Civilian coup leader Roberto Micheletti was named a congressman-for-life by the congress, a position that doesn't exist in the constitution. The National Industrialists' Association, a grouping of Honduras's most powerful businesspeople led by Miguel Facussé's cousin Adolfo Facussé, awarded Micheletti with their "National Hero" distinction, the first person to receive that honor in the 21st century. Military coup leader General Romeo Vásquez Velásquez was named by Lobo to head the state telephone company. In 2011, he announced his intention to run for president under the flag of the Patriotic Alliance, a political party he recently formed with other retired generals and colonels with a promise to bring order to Honduras.

In the middle of all this, the Lobo administration organized a massive conference called "Honduras is Open for Business." Potential investors were flown in from around the world to hear speeches on Honduras's new laws limiting labor rights and to review the hundreds of projects the government had on offer, from river concessions to a piece of Honduras's palm oil industry. Visitors were not informed that almost 10,000 hectares of those palm plantations were being occupied by campesinos, much less that the military was being deployed to threaten communities there. When Lobo himself was asked about the situation in the Aguán, he threw his arms in the air in

reiterate that it's a false solution to a global crisis. The problem is that the solution is always seen from a capitalist point of view and not from the point of view of the protection of human beings and the planet. Third World countries, the countries of Latin America and Africa, all of those countries that have the most resources, it's the most incredible thing: they have the most resources and the most vulnerable people. So right there, there is an ethical situation because the First World countries, which are the greatest consumers of resources are the ones that always set the rules about how to solve the problem."

disgust and said, "It's a beautiful day. We're talking about investment here. Everything is fine in the Aguán."

The UN Covers Its Ears

As news of the systematic assassination of farmers in the Aguán grew, a boycott and divestment campaign began that targeted Miguel Facussé. In 2010, Honduran human rights and social justice groups launched a boycott of Facussé's line of snack foods and cooking oils. In 2011, an international coalition of human rights, labor, religious, and food sovereignty organizations concluded a fact-finding mission to the Aguán, and then launched a divestment campaign. In response to the growing pressure, the German development bank (DEG) canceled a $20-million loan to Facussé in April of 2011. Three days after that, the world's largest utility company, France's EDF, announced it was canceling its contract to buy future carbon offsets from Facussé.

But the same week, the UK Secretary of Energy and Climate Change, Chris Huhne, responded to demands for the UK to withdraw its approval for the project by saying that responsibility for such determinations sat primarily with the Honduran government.

Then, in July 2011, the Executive Board of the UN's Clean Development Mechanism (CDM) gave final approval for Facussé's carbon credits. UK government climate negotiator and chair of the CDM Board, Martin Hession, defended the decision saying that it isn't the job of the CDM to screen for social concerns, adding that while they had received many complaints, the allegations weren't proven.

Dozens of campesinos are found murdered, often with signs of torture, on land that Facussé claims as his own. All in a period of mere months. Whose job is it to *prove* that Facussé is behind the attacks? The police often work together with Facussé's guards in evicting campesinos from land in conflict. In July 2011, it was the national police that burned to the ground the entire campesino community of Rigores. With torches and a bulldozer they destroyed three schools, two churches, acres of food crops, and the homes of more than 100 families. The police in the Aguán do not investigate Facussé. They work for him.

What about the journalists? Honduras, since Lobo took office, is now considered by Frank La Rue, the UN Special Rapporteur on Freedom of Expression, as "the most deadly country in the world for news gatherers". Like elsewhere, the Honduran media is dominated by a small group of families. Here they are all supporters of the coup and corporate control over land. The second largest newspaper in the country is owned by Facussé's nephew, former Honduran president Carlos Flores Facussé. Journalists with the courage to transgress editorial lines set by these publications live in constant danger. In fact, in the two years since Lobo took power, 19 journalists have been killed. While Lobo maintains that the deaths are unrelated to their work as journalists, the fact is that, of the 19 killed, 17 had published reports that challenged the coup regime.

One of the journalists killed was Aguán-based Nahúm Palacios. He was gunned down together with his girlfriend while driving home on March 14th, 2010. "They savagely assassinated Nahúm in order to shut him up," says Palacios's cameraman Mario Munguia, who fled the country after Palacios's murder. "They killed him just one week after we aired a series of reports on MUCA."

It was only after the Committee to Protect Journalists pressured the Lobo regime to investigate Palacios's murder that his grave was exhumed (three months after it was dug) in order to perform the autopsy the authorities had neglected to do in the first place. Unsurprisingly for Munguia, they never heard anything from the police, who have since dropped the investigation. "Their pseudo investigation provided zero results," he says, "either because they don't care, or because they know that the trail would end at some very powerful people."

Similar dangers await those within the campesino movement that attempt to document and disseminate their stories. Juan Chinchilla, who at just 26 years of age has become a key leader of MUCA, took on the task of collecting and distributing videos, photographs, interviews, legal documents, and other materials related to MUCA's struggle. In the two years since MUCA occupied the plantations, he has been jailed for 21 hours for not wearing a seatbelt, and threatened in person by both

uniformed police and armed men dressed in civilian clothing.

While he was taking pictures of the bodies of the five MCA members killed by Facussé's guards on November 15th, 2010, a police officer asked if he was the one uploading pictures to MUCA's website. He didn't answer, and another cop told him he would be next to die if he was.

He claims to receive regular death threats via text message. His friend Jorge received one during an interview for this book. It said: "We know you and little Juan are in Tegucigalpa. We're following you closely." We were, in fact, in Tegucigalpa.

The seriousness of such threats became clear on January 8th, 2011, when Chinchilla was kidnapped while returning home on his motorcycle. He still sports the scars from the open flame that they applied to his arms and head. He is also missing teeth from the beatings. After almost 24 hours of captivity, Chinchilla says he escaped while they were moving him through a remote path in the hills surrounding the Aguán Valley. He believes that if he hadn't escaped they would have surely killed him.

Chinchilla, who has lost dozens of friends in this struggle, remains a vocal leader of both MUCA and the youth commission of the FNRP. He is dedicated to ensure that nobody can claim that the atrocities taking place in the valley "aren't proven."

***

To produce such rich oil, palm trees need lots of sun, water, and nutrients. Their presence largely wipes out any competing species, thereby creating a monoculture that requires the mass application of pesticides to avoid a sweeping plague. The trees produce bunches of fruit weighing more than 60 pounds a piece, which are then boiled, spun, pressed, and filtered into vegetable oil, 70% of which is for export. For every tonne of fruit that enters the extraction plant, 0.065 tonnes of waste water are discharged into man-made lagoons. There it is treated to break down dangerous pollutants, a process that for decades has been releasing huge amounts of methane and CO_2 into the Aguán air. The resulting liquid is then dumped into a nearby river, eventually draining into the Caribbean Sea.

The money Facussé raises from CDM offsets will finance a biogas recovery system consisting of hoods to cover the slurry lagoons to collect the methane and convert it into heat and electricity to run the oil extraction plant. The plant currently runs on a mix of diesel, bunker fuel, and electricity taken from the grid. According to the official CDM report recommending the project for approval, it will save Facussé at least $350,000 per year in energy costs, on top of the millions he is now eligible to receive in carbon credits.

This discussion does not deal with the question of how effective these kinds of investments can be at mitigating the causes of climate change. For now, the UN and other gatekeepers of global capital flows have decided to go this route. Instead, we raise a different question: Who should benefit from this CDM-directed technology investment?

Prieta and the Clash of Models

It is surprising to some that the Honduran resistance should take its most organized and militant stand in the Aguán. It's not the poorest place in Honduras. Conditions are bad, but they are not at the borderline starvation levels seen in parts of Honduras's rural South. The Aguán, however, has one key distinction over other agricultural regions in the country — it's at once home to two models of farms that exist side-by-side. These are some of Honduras's largest corporate plantations, alongside the country's most successful campesino cooperatives. For campesinos, these co-ops serve as tangible, living alternatives, born almost 40 years ago out of organized struggle.

Co-operative communities like Prieta are made up of farmers just like any other. They put in similar work hours harvesting palm fruit and extracting its valuable oil. But unlike the workers on Miguel Facussé's plantations, Prieta members are their own bosses. And from that springs numerous advantages.

Workers on Miguel Facussé's plantations get no say in how their workplace is run, much less how the revenues of their work should be used. Members of Prieta form committees and make proposals on everything from budgeting to community safety. Their proposals are passed or defeated in general assemblies where members have equal voting rights. They've been working like this since 1974.

Workers on Miguel Facussé's palm plantations make roughly $3–4 per day. And any drop in global demand, adverse weather, or injury can lead to work layoffs. Meanwhile, members of the Prieta co-op bring in a guaranteed annual salary of $7,000, and are members for life, assuming they don't break certain mutually agreed-upon rules. Prieta President German Castro says with pride that in almost four decades, five members have been kicked out for hoarding or stealing.

Workers on Miguel Facussé's plantations are often unable to send their kids to school. Not only do all children of Prieta's members attend schools in the community, the co-op provides full scholarships to any student that is accepted to college or university. "Five percent of our revenue," explains Castro, "is dedicated to education. We're already seeing the benefits of our education policy. In the next generation, we already have lawyers, doctors, economists, civil engineers, a whole series of people with diverse skills and a dedication to the co-op. This is our strength. They are going to have the new ideas, the new creativity that will keep this co-op very strong. We are proud that our co-op is a model for Honduras and all of Latin America."

But, their model co-op is under attack. In February 2011, then co-op President Rigoberto Fúnez, a local legend for his leadership in Prieta's difficult early years, was assassinated while driving on the highway—in plain daylight. Also killed in the attack was Prieta's Treasurer Fredy Castro, German's brother. German Castro was elected to replace Fúnez. The co-op voted to hire independent investigators to look into the killings after the police appeared unwilling to investigate political motivations.

Fúnez was also the president of the Prieta chapter of the FNRP, which has been a key supporter of local campesino movements like MUCA. "Before the coup, we were indifferent to the problems of other campesinos. We couldn't care less if they formed their own co-ops or not,"

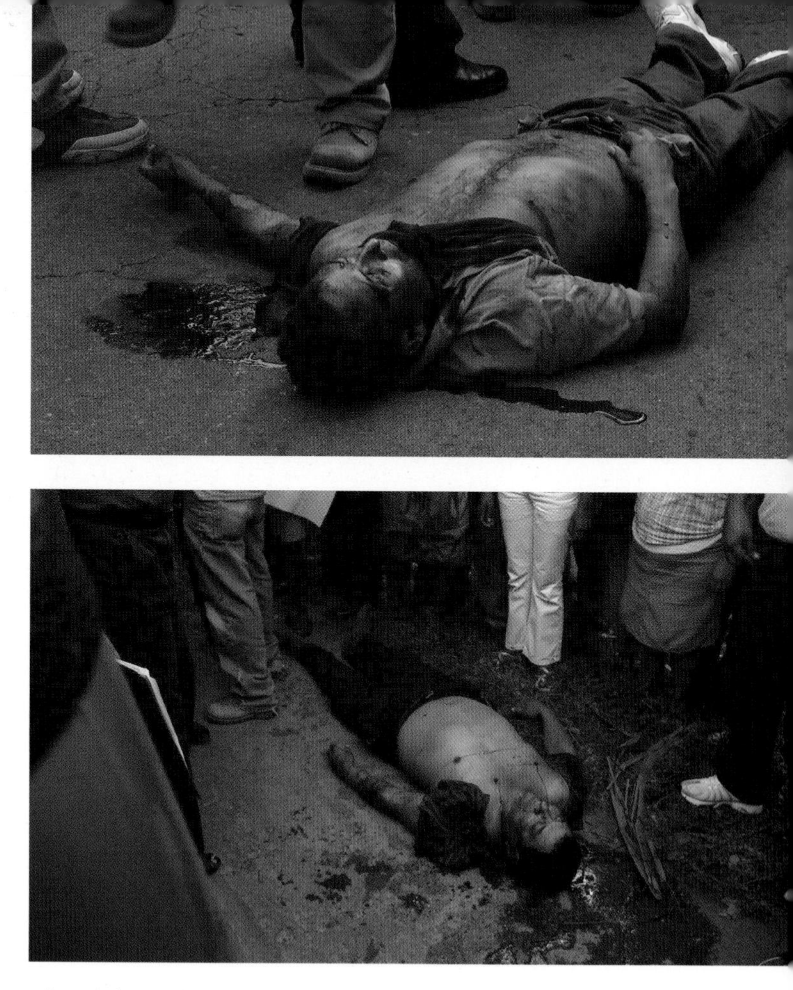

Above: On February 10th, 2011, two hours before a scheduled interview for *The Carbon Rush*, Rigoberto Fúnez and Fredy Gonzalez Castro, the President and the Treasurer of the Prieta Palm Cooperative, were assassinated. Mario Mumbreno, a community organizer based in Tocoa explained that, "Undoubtedly, the deaths of the two comrades were part of the strategy against the National Front of Popular Resistance in this region.

says Castro. "Thank God for the coup! It took everyone's mask off. We realize now that we all need to be unified because their goal is to impose their system everywhere, even here in Prieta."

In September 2011, after these interviews, German Castro and his wife Miriam Emelda Fiallo were gunned down while driving. Emelda Fiallo died from her wounds. Castro barely survived and today lives in exile, paralyzed from the waist down.

***

Over the next ten years Facussé will receive somewhere between $2-3 million from selling offset credits in Europe. It's not a game changer, especially when compared to the $30 million the World Bank recently invested in Facussé's Dinant Corporation. The CDM decision does, however, provide one of the most absurd examples of how the "international community" routinely sides against the organized poor in Honduras.

"We're not against foreign investment," explains Prieta President Castro, "We just want investors to be held to the laws that we will set out in a new constitution." The UN's decisions in the Aguán Valley have demonstrated why Honduran campesinos demand a say in the decisions made by foreign investors.

The UN essentially showed up in the Aguán Valley and announced, "I have a

new piece of technology (biogas capture) that will save you millions of dollars on your costly energy needs while reducing the ecological impact of your activities." From the top of Mount Botaderos, overlooking the sea of palm trees covering the valley, they observed two models of palm oil production—the co-op model and the corporate model—that could benefit from their gift.

The co-op model offers each of its members an equal say in running of the enterprise. Over the years, their families have achieved a dignified salary, health care, education, job security, and university scholarships. Co-ops see the poor and landless farmers around the valley as sisters and brothers, doing what they can to help other campesinos get the land, advice, and technical assistance they need to form their own co-ops and live with dignity.

The corporate model offers its employees poverty wages, frequent layoffs, and zero benefits. Instead of investing its revenues on health care, education, and wages, it supports military coups and hires paramilitaries to stop the poor and landless farmers from getting the land, advice, and technical assistance they need to form co-ops and live dignified lives.

The UN chose the corporate model. The campesinos of the Aguán continue to give their lives, lest they be reduced to laborers for a slightly greener *golpista* (coup détat).

FNRP
SUPERACION
RMAS

The Colonization Aguán

For decades, many Latin Americans have argued that to progress out of poverty and its indignities requires true agrarian reform. The unequal economic and political influence enjoyed by small groups in countries like Honduras is directly tied to their control over the most fertile soils. In 1959, the Cuban Revolution showed that land could be expropriated and made into campesino co-ops—just 90 miles from the shores of US Campesinos around the continent began organizing to do the same. Soon US President John F. Kennedy was advised to do something to alleviate the growing revolutionary pressures throughout rural Latin America.

In an essay published by Princeton University, Mexican scholar Edmundo Flores advised Kennedy that there were only two ways for Latin America to democratize: by taxing the elites and investing in the poor or by taking the elites' land and giving it to the poor. Flores concluded that the first was impossible, as no group of elites would agree to tax themselves. So, he reasoned, if Kennedy hoped to aid in the democratization of Latin America while also avoiding a wave of anti-imperial revolutions, he would have to break the US's traditional alliance with the hemisphere's elites, and support those groups working toward expropriation and agrarian reform.

On March 13, 1961, Kennedy announced the creation of the Alliance for Progress, a "Marshall Plan for Latin America," to focus in large part on an agrarian reform that would "make land ownership a possibility" for the rural poor. Two days before his address, Kennedy had secretly given his approval for the CIA-led invasion of Cuba. In April, a group of Cuban exiles trained and equipped by the US was soundly defeated by the Cuban military at the Bay of Pigs. Revolutionaries around Latin America celebrated another loss for

People's resistance is often met with brutal force.

Security is tight; conflict is not uncommon

the US. The Alliance for Progress would be more necessary than ever if the US wanted to maintain its dominance in Latin America. In August, at the official launch of the Alliance, Kennedy pledged a whopping $20 billion over ten years.

Flores's essay warned the US that:

> *...the Alliance will have no other choice, at the "moment of truth," than to oppose or to favor revolutionary change. If, following current misconceptions, the US backs the quasi-feudal and militaristic governments in power, there will be a pretense of economic development and Alliance funds will be misallocated and wasted without changing the conditions responsible for political unrest and economic stagnation. This will lead eventually to the establishment of military dictatorships of the extreme right.*
>
> *History would unfold exactly as Flores predicted. The "quasi-feudal governments" went on to funnel the Alliance money to the existing plantation system, decisions that went unopposed by US bureaucrats. Governments that attempted true agrarian reform by expropriating land, typically unused land, from the wealthiest and opening it to the landless, were overthrown by militaries trained and supported by the US. Such occurred in Brazil in 1964 and Chile in 1973. Weak land reform laws were passed in 15 Latin American countries in the 1960s alone, but they either were never implemented or took the form of colonization instead of expropriation. This is how the Aguán came to be.*

Honduras was seen as a likely place for a leftist revolution. Workers toiled in feudal conditions with no influence on politics. Most importantly, however, Hondurans had already witnessed the potential force of collective action. Honduras's 1954 general strike virtually shut down the country's banana industry for 69 days, at the time of the country's largest export, winning wage increases and union recognition. So, to avoid that kind of bottom-up political change that

was being threatened around the country, the military dictatorship joined with the US Agency for International Development (USAID) in opening up the Aguán Valley for colonization. Campesinos flocked from around the country to clear the virgin jungle and received land titles. Seventy campesino co-ops were created, under the sole condition that they plant African palm trees.

Once the co-ops reached productivity, the government began to change the game. In the early 1990s, President Leonardo Callejas implemented, as a condition of World Bank and IMF loans, sweeping economic reforms that included devaluing the lempira (Honduran currency), lifting agricultural tariffs, and government price guarantees. All were rule changes that favored large corporations over small co-ops. Then in 1992, Callejas passed the Agricultural Modernization Law, reversing the Aguán's special status as a land reform zone, removing the 300 hectare limit on land holdings and allowing co-ops to sell their lands to corporations. The sales that followed were helped along by misinformation, bribes, and intimidation. "They were offering each campesino 6,000 lempiras and telling them they could spend the rest of their lives in a hammock living off the interest. To men who had never seen 1,000 lempiras before! Then, and we all remember this, they devalued the currency right afterwards." In just two years, over two thirds of the land originally titled to campesino cooperatives had been sold to large plantation owners.

Community under Construction

The new generation of groups like MUCA are led mostly by campesinos who were too young to vote in 1992. Juan Chinchilla was only six years old. They feel that the original land reform is more legitimate than the 1992 Callejas law, and that even if it isn't, the success of groups like Prieta shows the superiority of the co-op model.

And they believe that if the government won't apply it then they'll do it themselves. In fact, they believe that if any president they vote in to change the course is going to be overthrown, then they'll occupy the plantations and impose the model by force. In these conditions, MUCA is implementing lots of new ideas.

While the old co-ops managed to free themselves from the bosses at work, the boss at home remains. At the Salama cooperative, for example, women have no vote. At the Prieta co-op, five women are now voting members, only because their husbands or fathers passed away, and they are outnumbered 7 to 1 by their male counterparts. Prieta members concede openly that patriarchy is a serious weakness in the movement, one that they are only now recognizing since the coup. Their participation in the FNRP and sharing committees with feminist organizations has caused them to reflect on many things.

But, in the new occupations under MUCA, women are often leading the way. "If women could have voted in 1992, they would have never sold to Facussé," says Orvelina, one of MUCA's eldest members. "I'm a proud co-founder of MUCA, and we women are playing a big role this time around, which is great because if we're gonna succeed, we're gonna need everybody."

Food vs. Fuel

Women's participation isn't the only progress taking place in occupied territory.

Despite being the country's most fertile valley, markets in the Aguán carry only imported rice, mostly from the US, at prices many times higher than local rice was sold here just a decade ago.

First generation co-ops like Prieta are only now beginning to plant other crops, after calculating the outrageous percentage of revenues spent on basic food stuffs. If their first mistake was excluding women, they say, their second was focusing exclusively on palm production.

Second generation occupations like MCA made some strides. MCA produces all its own milk, with every child getting a liter per day, but they are still at the whim of global food markets when it comes to basic grains and vegetables.

But the third generation occupations, those that established themselves after the coup, have made huge advances in just two years. At MUCA, they have begun raising poultry, built a fish farm, planted various crops including corn and watermelon, and finished a new three-story building to house a consumer co-op, meeting hall, and eventually their own radio station.

While the CDM documents include detailed estimations on how much carbon emissions will be saved through Facussé's biogas plant, nowhere in these calculations will one find any mention of the emissions created by shipping rice from Arizona to the Aguán.

THE RUSH IN PARANA, BRAZIL

by Antonio Cabrera Tupa
Vice President of OSCIP Guarany, the largest organization working for the rights of indigenous Guarani peoples

The Guarani are an indigenous people populating the triple frontier of Brazil, Paraguay, and Argentina. They have been here for millennia but only recently has big business from the North taken any interest in them. Well, not the Guarani, but their land, particularly the land around Iguaçu Falls, in the state of Paraná, Brazil.

The Guarani in the region have been custodians of the land for as long as anyone can remember, but now the government here has termed them nomadic and, as such, has deemed them a landless group. When the land is not "owned" it is easier to take it over. And that is what has been done here in the name of carbon offsets in the form of the Guaraqueçaba project, run by the Nature Conservancy and the Society for Wildlife Research and Environmental Education. Northern firms invest in green businesses here which in turn take over the people's land (the means of their sustenance and cultural continuity). Why? To offset pollution quotas in the North so big business can continue to pollute the sky and waters up there. As a result, the Guarani suffer territorial displacement, racial segregation, harassment, and worse, all in the name of land acquisition and the maintenance of those "green" businesses that invade Guarani territory. As we've seen in other parts of the world,

The water landing to Tekoha Kuarahy a Guarani village.

The conservation project covers a vast area including the traditional land of the indigenous Guarani people.

"It's a forgotten village, abandoned by the municipal and federal governments. They've been abandoned. The biggest concern is that this land they live on is theirs. They have a right to it. They have the right to stay here, but they don't own it."
Michelle do Pilar Lemos, local teacher

Karai Jeguaka Verá, Guarani Spiritual Leader

these industries do little to protect—let alone—respect local populations.

Here, in Brazil, the results have been shocking and disturbing. The Guarani, many of whom are displaced, become homeless and destitute and turn to drugs and prostitution. Their culture and beliefs stem from Mother Earth and without it, they lose their stories, their culture, their hope. Aboriginal peoples know this story well; it's been happening ever since Christopher Columbus and it continues to this day, just in a different guise.

The cast of carbon credit characters that are destroying the people here are many, representing big business and non-governmental agencies alike. They are to blame of course, as are the governments here that allow such things to happen to their own people. But these businesses and these governments find legitimacy and support in the United Nations' mechanisms that first created this crazy carbon rush for profits. The result is that the UN is lending "credence" and "credibility" to the ongoing push to privatize indigenous land, here, and elsewhere in the South. This is dead wrong, and these Kyoto carbon offsetting schemes need to be overturned, before the Guarani in Brazil become another cultural footnote in the North's quest for profits.

Karai Jeguaka Verá preparing food from the palm tree he cut down. Heart of palm is the most common food for the community. Within the place now being considered a conservation zone, they are no longer allowed to cut down palm trees for any purpose.

The Força Verde is the environmental police of the state of Parana, protecting the conservation project.

Karai Jeguaka put the reality of his community simply: "They don't want us to enter the land in which we have lived before. We didn't give them the land. They want to take our home from us."

Karai Jeguaka Verá is an 81-year-old Guarani shaman. He knows the forest well and as he put it, “When my children get sick I prepare medicine to cure them”. He is seen preparing heart of palm soup.

Everyday the Força Verde go out on patrol to make sure people are not squatting on land in the conservation territory.

Karai Jeguaka Verá spells out the hardships of his community: "We barely have food. We can't hunt or harvest wood. They prohibit us. But there are other people who enter the forest and steal the wood. And then they come to ask who did it. We go and cry in the forest. What are we going to do?"

1. Patrols often consider local people squatters, and regularly push locals off their traditional land.

2. Guarani Chief Valdinei Verá da Silva explains, "All our lands have been taken away from us. We want them back. We have been abandoned with no land."

3. Juliano Bueno De Araujo of the Aninpa Institute explains how "Hunger is now common, and there's endless social problems. These conservation projects are to protect the forest and capture carbon to keep the planet from warming, but you have entire communities dying of hunger. It's unacceptable."

4. The Força Verde also regularly patrol the ocean bay within the conservation zone. Instead of focusing on ensuring that large scale fishing is not occurring, small fisherman are regularly harassed and surveilled.

"The Guarani people are not well. We live in poor conditions, and nobody knows about it. We hardly get by. We do not benefit from the conservation project. We see no future."

CONCLUSION

CARBON RUSH OR CLIMATE JUSTICE?

By Patrick Bond, political economist and activist who directs the Centre for Civil Society at the University of KwaZulu-Natal in Durban, South Africa

"What happens when we manipulate markets to solve the climate crisis? Who stands to gain and who stands to suffer?" Amy Miller's questions are profound and it strikes me that only the climate justice perspective begins to provide a satisfying answer.

Though carbon markets were theorized from the late 1980s and made the core of global climate policy in 1997, this critical perspective was evident to me only a decade ago, after being introduced to Sajida Khan (1952-2007). Khan was a lifelong Durban resident whose fight against a Clean Development Mechanism (CDM) project—a methane-electricity generator in Africa's biggest landfill - was seminal for many climate justice activists' understanding of carbon markets. Having fought for the landfill's closure since it opened across from her home in 1980, she realized in 2002 that the World Bank's CDM financing offer would keep the Bisasar Road dump open for many more years. Along with many in the progressive movement, we learned of her cause just before the 2002 World Summit on Sustainable Development, when the organizations groundWork and CorpWatch invited Khan to share the Durban story at a conference of activists. That led to the TransNational Institute's 2002 film *Green Gold* (and in 2005 two others by South African filmmakers Rehana Dada and Aoibheann O'Sullivan) which gave more people access to the complicated debate about carbon trading in both macro and micro terms. (In 2009, another film, *Story of Cap and Trade* by Annie Leonard, was seen by more than a million people.)

Khan's struggle against both the landfill and carbon trading made the front page of the *Washington Post* in March 2005, just as the Kyoto Protocol came into effect.[1] The Bank, which had catalyzed and still enthusiastically backed Durban's CDM application, suddenly took fright when she filed a 70-page environmental impact lawsuit in local courts to block the project, in the process generating global support for her struggle. Though the project went ahead through private carbon markets, that retreat was a small victory but an important reminder of how determination, technical sophistication, and community support reminiscent of Erin Brokovich can intimidate power.

To be sure, Khan didn't get the support of everyone nearby: the Kennedy Road shackdwellers living on the edge of the landfill, whom the Bank and its municipal allies promised jobs and bursaries, demonstrated against her: "There are comrades who are saying, 'Bush take your millions away.' How can Bush take his millions away?" asked Sbu' Zikode in 2005.[2] Three years later, Zikode —an eloquent leader of one of South Africa's most important 21st century social movements, Abahlali baseMjondolo, acknowledged that Durban municipal officials manipulated the Bisasar Road area's socio-racial divisions: "We were used. They even offered us free busses to protest in favor of this project ... to damage those who oppose this project."[3] The promised jobs and bursaries never materialized, and a fancy truck acquired by another Abahlali leader led to more internecine conflict. That too was a lesson in divide-and-conquer politics associated with the CDM funding offered to impoverished people (even when extremely well organized, like Abahlali), a lesson taught repeatedly in the carbon and offset markets, and especially the forestry (REDD) schemes that later divided indigenous peoples.

These are not minor localized problems, for the stakes could not be higher. In 2006, Christian Aid estimated that 182 million Africans were at risk of premature death due to climate change this century.[4] In 2009, former UN secretary general Kofi Annan's Global Humanitarian Forum issued a report, "The Anatomy of a Silent Crisis," which provided startling estimates of damages already being experienced:

> *An estimated 325 million people are seriously affected by climate change every year. This estimate is derived by attributing a 40 percent proportion of the increase in the number of weather-related disasters from 1980 to current to climate change and a 4 percent proportion of the total seriously affected by environmental degradation based on negative health outcomes ... Application of this proportion projects that more than 300,000 die due to climate change every year.*[5]

What can be done to prevent this? The answer from the climate justice movement —drawing upon April 2010 Cochabamba, Bolivia conference declarations—includes the decommissioning of carbon markets and the CDM mechanism and their replacement with a suitable climate debt payment system. (Such a system would directly channel resources to climate

1 *Washington Post*, "Kyoto credits system aids the rich, some say," 11 March, 2005, http://www.washingtonpost.com/ac2/wp-dyn/A28191-2005Mar11?language=printer

2. Aoibheann O'Sullivan, "Carbon credits and Durban's Dump", *CCS Wired*, 2006, http://ccs.ukzn.ac.za and http://www.aoibheann.net/film2005.html

3. groundWork, *Wasting the nation*, 2008, http://www.groundwork.org.za/Publications/gWReport2008.pdf

4. http://www.christianaid.org.uk/pressoffice/pressreleases/august2008/climate_change_talks_accra.aspx

5. Global Humanitarian Forum, "The human impact of climate change," New York, 2009, http://www.global-humanitarian-climate-forum.com/uploads/An_Impacts.pdf.

victims without corrupt aid-agency and middlemen or venal state elites.)

Instead, global climate governance by elites continues to make matters worse. Each year, the United Nations Framework Convention on Climate Change Conference of the Parties (COP) meets to deliberate on a framework for global emissions cuts and crisis-adaptation strategies. In December 2011, Durban hosted the COP 17, and delegates once again heard that the solution to climate crisis must center on markets, in order to "price pollution" and simultaneously cut the costs associated with mitigating greenhouse gases. Moreover, say proponents, these markets are vital for funding not only innovative carbon-cutting projects in Africa, but also for supplying a future guaranteed revenue stream to the Green Climate Fund (GCF). That fund was meant to contain $100 billion per year for spending by 2020 (according to a promise by US Secretary of State Hillary Clinton at the 2009 Copenhagen COP 15), but Durban's COP 17 set it up merely as an "empty shell," since only a few of the rich countries pledged funds. Sensing that the promise would be broken, GCF design team co-chair Trevor Manuel (South Africa's Planning Minister) argued as early as November 2010 that up to half GCF revenues would logically flow from carbon markets.

***

The carbon market strategy was established within the Kyoto Protocol in 1997. It aims to facilitate innovative carbon-mitigation and alternative development projects by drawing in funds from northern greenhouse gas emitters; in exchange, those northern countries can continue to polluting. CDMs, like the Bisasar Road landfill, generate Certified Emissions Reductions (CERs) that act as another asset class to be bought, sold, and hedged in the market. The European Emissions Trading Scheme (ETS) is the main trading site.

CDMs were created within Kyoto to allow wealthier countries classified as "industrialized"—or Annex 1—to engage in emission reduction initiatives in poor and middle-income countries, as a way of avoiding direct emissions reductions. Put simply: the owner of a major polluting vehicle, like Shell, can pay an African country to not pollute in some way, in order that Shell can continue with its emissions. In the process, developing countries are, in theory, benefitting from investment in sustainable energy projects. The use of such "market solutions to market problems" will, supporters argue, lower the business costs of transitioning to a post-carbon world. After a cap is placed on total emissions, the idea is that high-polluting corporations and governments can buy ever more costly carbon permits from those polluters who don't need so many, or from those willing to part with the permits for a higher price than the profits they make in high-pollution production, energy-generation, agriculture, consumption, disposal or transport.

The market is meant to incentivize greenhouse gas emissions cuts, which scientists say should be 50 percent of Northern emissions by 2020. Ideally, the GCF would also ensure that the North's climate debt to the South covers the sorts of damages Annan specified under a "polluter pays" logic, or for establishing a transition path to a post-carbon society and economy. What chance is there for these mechanisms to work as desired?

The first decade looked promising. With Europe as the base, the world emissions trade grew to around $140 billion in 2008 and was at one point projected to expand to $3 trillion/year by 2020 if the US signed on. The $3 trillion estimate didn't even include the danger of a bubbling derivatives market, which might have boosted the figure by a factor of five or more.[6] But the markets have been extremely volatile and began trending downwards in 2006, when the price rose to $40/tonne. It subsequently crashed to less than $10/tonne due to economic meltdown, increasing corruption investigations within the ETS, and COP-induced despondency.

Recent evidence of market efficacy is damning. Even within the very limited, flawed strategy of carbon markets, there were mixed outcomes from the Durban COP 17. In spite of Trevor Manuel's efforts to bring emissions trading into the GCF, where it does not belong, and in spite of the United Nations CDM Executive Board's decision to allow "Carbon Capture and Storage" experiments to qualify for funding, the most profound flaws in the existing market were not addressed. Without an ever-lowering cap on emissions, the incentive to increase prices and raise trading volumes disappears.

Worse, in this context of economic stagnation, financial volatility, and shrinking demand for emissions reduction credits, the world faces increasing sources of carbon credit supply in an already glutted market. And fraud continues, including in Durban's own celebrated pilot CDM project, the Bisasar Road landfill.

Because the Durban COP 17 left the world's stuttering carbon markets without a renewed framework for a global emissions trading scheme, the emissions trade then crashed even farther, suffering a 20 percent decline in the first three months of 2012. Durban turned the Kyoto Protocol—which is now applicable to only 14 percent of world greenhouse gas emissions—into

6. See Nina Chestney and Michael Szabo, "Emissions traders expect US carbon market soon," Reuters, May 28, 2009, http://www.reuters.com/article/GCA-GreenBusiness/idUSTRE54R4YP20090528, last accessed October 11, 2009.

a "Zombie" (walking-dead) because its heart, soul, and brain (binding emissions cuts) all died, as former Bolivian ambassador Pablo Solon put it.[7] All that appears to be moving is the stumbling and indeed crashing commitment to CDMs.

These markets can be expected to die completely if Qatar's COP 18 does not generate more commitments to legally-binding emissions cuts. And judging by Washington's threat, it won't be until 2020—the COP 26!—that the United States will review its own targets: the Copenhagen Accord's meaningless 3 percent cuts offered from 1990–2020. By then it will be too late, because the Kyoto Protocol's mistaken reliance on financial markets means that the period 1997-2011 will be seen as the lost years of inaction and misguided financial quackery—when we urgently need the period going forward from 2012 to be defined as the era when humanity finally took charge of its future and ensured planetary survival.

For those hoping Durban would provide a better global-scale negotiating terrain, the opportunity has been lost. The balance of forces will not improve in Qatar in December 2012, given the prevalence of irresponsible major powers—best represented by Ottawa's withdrawal from the Kyoto Protocol just after the COP 17—and the probability that in Washington, Republican Party rightwing climate deniers will prevent further concessions. There are no prospects that the European Union's Emissions Trading Scheme will turn around in the near future, and only a few minor national and subnational trading experiments appear on the horizon. Only the $100 million World Bank-European Union "Partnership for Market Readiness" continues the myth that markets are an appropriate strategy, through grants to gullible officials in Chile, China, Colombia, Costa Rica, Indonesia, Mexico, Thailand, Turkey and Ukraine. As even the pro-trading Point Carbon news services remarked just after the Durban COP 17 ended:

> *Such initiatives are essential to ensure new markets get off the drawing board because a nervous private sector has little appetite to invest in new programmes without further political guarantees that someone will buy the resulting credits.*[8]
>
> *Reuters confirmed the analysis, saying that carbon markets are on "life support" and the COP 17 failed to deliver "a needed boost in carbon permit demand."*[9]

The EU system was meant to generate a cap on emissions and a steady 1.74 percent annual reduction, but the speculative character of carbon markets gave perverse incentives to stockpile credits. Large corporations as well as governments like Russia gambled that the price would increase from low levels to double or triple the prices (as promoters continually predicted). Instead, now, with the market collapsing, the next perverse incentive is to flood the market so as to at least get some return rather than none at all when eventually the markets

7. Pablo Solon, Wolpe Lecture at the University of Kwa-Zulu-Natal, Durban, 2 December, 2011, http://ccs.ukzn.ac.za.

8. Susanna Twidale, "Durban deal delays debate on new markets", *Point Carbon,* 13 December, 2011.

9. Reuters, "Carbon markets still on life support after climate deal", 13 December, 2011.

are decommissioned. This is precisely what happened to the Chicago climate exchange—the US carbon market—in late 2010, and many of those who held shares in the exchange subsequently sued the high-profile founder, Richard Sandor, for misrepresenting the value of their assets, a strategy that should repeat across the world given the prolific false claims associated with carbon markets.

As a result, no investor believes there is any money to be made by utilizing carbon markets to direct climate-conscious investments. A month after Durban's denouement, it was evident to the French bank Société Générale that "European carbon permits may fall close to zero should regulators fail to set tight enough limits in the market after 2020", and without much prospect of that, the bank lowered its 2012 forecasts by 28 percent.[10] The 54 %crash for December 2011 carbon futures sent the price to a record low, just over €6.3/tonne.

Worse, an additional oversupply of 879 million tonnes was anticipated for the period 2008-2020, partly as a result of a huge inflow of UN offsets: an estimated 1.75 billion tonnes. This glutting problem is not only due to the demand deficit thanks to the COP 17 negotiators' failure to mandate emissions cuts, but is also in part due to the lax system the UN appears to have adopted. All manner of inappropriate projects appear to be gaining approval. According to Professor David Victor, a leading carbon market analyst at Stanford University, as many as two-thirds of registered carbon emissions reductions do not constitute real cuts.[11]

In 2004, a global civil society network, the Durban Group[12], was formed to oppose carbon trading. From the vantage point of an austere Catholic mission on Durban's highest central hill, the Glenmore Pastoral Centre, a score of the world's critical thinkers were brought together by the Swedish Dag Hammarskjold Foundation. They deliberated over the neoliberal climate fix for several days. We worried that the main test case, the EU's Emissions Trading Scheme, not only failed to reduce net greenhouse gases there, but suffered extreme volatility, an inadequate price, the potential for fraud and corruption, and the likelihood of the market crowding out other, more appropriate strategies for addressing the climate crisis. The critique can be summed up in eight points:

- the idea of inventing a property right to pollute—by selling permits to emit to the highest bidder—is effectively the "privatization of the air," a moral problem given the vast and growing differentials in wealth inequalities
- greenhouse gases are complex and their rising production creates a disjointed impact—the escalation of climate change after a tipping point—which cannot be reduced to a commodity exchange relationship (a tonne of CO_2 produced in one place is accommodated by reducing a tonne in another, as is the premise of the emissions trade)

10. Catherine Airlie and Matthew Carr, "EU, UN carbon prices could fall 'close to zero', SocGen says", Bloomberg, 17 January, 2012.

11. John Vidal, "Billions wasted on UN Climate Programme," *The Guardian,* 26 May, 2008, http://www.guardian.co.uk/environment/2008/may/26/climatechange.greenpolitics

12. http://www.durbanclimatejustice.org/

- the corporations most guilty of pollution and the World Bank—which is most responsible for fossil fuel financing—are the driving forces behind the market, and can be expected to engage in systemic corruption to attract money into the market even if this prevents genuine emissions reductions

- many of the offsetting projects—such as monocultural timber plantations, forest "protection" and landfill methane-electricity projects—have devastating impacts on local communities and ecologies, and have been hotly contested in part because the carbon sequestered is far more temporary (since trees die) than the carbon emitted

- the price of carbon determined in these markets is haywire, having crashed by half in a short period in April 2006 and by two-thirds in 2008, by another 50% during 2011, thus making mockery of the idea that there will be an effective market mechanism to make renewable energy a cost-effective investment

- there is serious potential for carbon markets to become an out-of-control, multi-trillion dollar speculative bubble, similar to exotic financial instruments associated with Enron's 2002 collapse (indeed, many former Enron employees populate the carbon markets)

- as a "false solution" to climate change, carbon trading encourages merely small, incremental shifts, and thus distracts us from a wide range of radical changes we need to make in materials extraction, production, distribution, consumption, and disposal

- the neoliberal ideology of finding market solutions to market failures rarely makes sense, and that ideology is in spectacular disrepute following the world's worst-ever financial market failure, largely because the very idea of derivatives—a financial asset whose underlying value (e.g. the right to pollute) is several degrees removed and also subject to extreme variability—was thrown into question.[13]

In short, the return of market mania to climate negotiations is a dangerous diversion from a daunting reality: that the US, China, South Africa, and most other big emitters want to *avoid* making the binding commitments that are required to limit the planet's temperature rise, ideally below the 1.5°C that scientists insist upon this century. Naturally, the (binding) Kyoto Protocol is a threat to the main emitting countries, which have been working hard since early 2010 to replace it with the voluntary, loophole-ridden Copenhagen Accord. This is the easiest

13. The analysis of emission market contradictions generated by Larry Lohmann is probably the most sophisticated, e.g., see Lohmann, L. 2006. "Carbon trading: A critical conversation on climate change, privatisation and power," *Development Dialogue*, 48, September, http://www.dhf.uu.se/pdffiler/DD2006_48_carbon_trading/carbon_trading_web_HQ.pdf; Lohmann, L. (2009a), "Climate as Investment", http://www.thecornerhouse.org.uk/pdf/document/Climate%20as%20Investment.pdLohmann, L. (2009b), "Neoliberalism and the calculable world: The rise of carbon trading", in K. Birch, Mykhnenko, V. and Trebeck, K. (eds.), *The Rise and Fall of Neoliberalism: The Collapse of an Economic Order?*, London: Zed Books; Lohmann, L. (2009c), "Regulatory challenges for financial and carbon markets," in *Carbon & Climate Law Review*, 3(2); Lohmann, L. (2009d), "Toward a different debate in environmental accounting: The cases of carbon and cost-benefit", in *Accounting, Organisations and Society*, 34(3-4): 499–534; and Lohmann, L. (2010), "Uncertainty markets and carbon markets: Variations on Polanyian themes", http://www.thecornerhouse.org.uk/pdf/document/NPE-2high.pdf

way to understand the procrastination and lack of ambition in the December 2011 Durban deal.

And naturally, the North's failure to account for its vast "climate debt" continues. To illustrate, Pakistan suffered $50 billion in climate-related flood damage alone in 2010, yet the total on offer from the North to the whole world was just $30 billion for 2010–12, according to promises made in Copenhagen. By the time of the Durban COP 17, there was no realistic chance that $30 billion in North-South flows would actually be delivered.

Climate negotiators should have known that carbon trading was a charade that would do nothing to reduce global warming. What was meant as an incentive scheme to provide stability and security to clean energy investors had become the opposite. A low and indeed collapsing carbon price—€6.70/tonne in July 2012, down from a peak five times higher six years earlier—was useless for stimulating the kind of investment in alternatives needed: for example, an estimated €50/tonne (at minimum) is required to activate private sector investments in "carbon capture and storage," the as-yet-non-existent (and extremely dangerous) technology by which coal-fired power stations could, theoretically, bury liquefied carbon emitted during power generation. Substantial solar, tidal, and wind investments would cost much more yet. The extreme volatility associated with emissions trading so far makes it abundantly clear that market forces cannot be expected to discipline polluters.

The only real winners in emissions markets have been speculators, financiers, consultants (including some in the NGO scene), and energy sector hucksters who made billions of dollars in profits on the sale of notional emissions reduction credits. As the air itself became privatized

and commodified, poor communities across the world suffered and resources and energy were diverted away from real solutions. But one of the most powerful set of critiques came from the inside: internal contradictions which repeatedly crashed the market and prevent it from carrying out actual emissions reductions.

These problems were sensed, to some extent, by the very founders of the environmental markets. Canadian economist John Dales (who died in 2007) first justified trading in emissions rights by applying market logic to water pollution in a seminal 1968 essay, "Pollution, Property, and Prices." Waste quotas were imposed along with a market in "transferable property rights ... for the disposal of wastes," interchangeable among firms.[14] Thirty-three years later, he expressed doubts about carbon markets in an interview: "It isn't a cure-all for everything. There are lots of situations that don't apply. It is not clear to me how you would enforce a permit system internationally. There are no institutions right now that have that power."[15] Also in the late 1960s, in the US, graduate economics student Thomas Crocker had famously advocated emissions trading for discrete problems, but in 2009 he told The Wall Street Journal, "I'm skeptical that cap-and-trade is the most effective way to go about regulating carbon."[16]

If so, we need to return to the opening queries, "What happens when we manipulate markets to solve the climate crisis? Who stands to gain and who stands to suffer?" The winners are polluters who buy carbon so they can keep emitting greenhouse gases; companies that sell carbon often in whimsical and unethical ways; and carbon speculators. The losers are not only those like Khan and Kennedy Road residents, but the vulnerable populations of Africa, the Andes, the Himalayas and all those who depend on the mountain spring water; and small islands and all those living in places vulnerable to extreme weather events, not excluding those in wealthy countries, in places like New Orleans whose poorest citizens were left devastated. The winners within the Third World are a few countries' governments and capitalists. The vast bulk of financing has gone to just four countries: China, India, Brazil, and Mexico (together issuing more than two thirds of CDM credits). The short answer is that the 1 percent at the top of the socio-economic pyramid wins, and everyone else loses. And this calculus deserves the attention of all of us trying to change that kind of destructive power wherever we find it.

14. Dales, J. 1968. *Pollution, Property and Prices: An Essay in Policy-Making and Economics.* Toronto: University of Toronto Press, p.85.

15. Jon Hilsenrath, "Cap-and-trade's unlikely critics: Its creators—economists behind original concept question the system's large-scale usefulness, and recommend emissions taxes instead," *Wall Street Journal*, August 13, 2009.

16. *Ibid.*

THE CARBON RUSH
THE CARBON RUSH
THE CARBON RUSH

FIVE THINGS YOU CAN DO TO STOP EMISSIONS TRADING

1. Support Indigenous Struggles

In 2009, Indigenous communities throughout the world called for a global mobilization. "In Defence of Mother Earth" on October 12, 2010, was created as a result to reclaim "Columbus Day" and transform other colonial celebrations into days of action and solidarity with Indigenous peoples.

Resistance by Indigenous community to pressures from large corporate entities represents an act of protection, and embodies visceral alternatives for sane resource management in Haida Gwaii, for conservation of watersheds in Gwich'in, for sustainable forestry in Barriere Lake, and for imagining different environmental relationships from coast to coast to coast. Where polluting and carbon-emitting projects have been halted or delayed, minimized or regulated, we can thank Indigenous peoples for their determination in implementing real change. During these struggles, Indigenous communities have forged a powerful set of tools, including Supreme Court precedents, constitutional rights, and international legal instruments that establish a framework for self-determination and land restitution in Canada.

2. Shut Down the Oil Sands

The Oil Sands "Gigaproject" is the largest industrial project in human history, and arguably the most destructive. Oil Sands mining procedures release at least three times as much CO_2 emissions as regular oil production, yet the Oil Sands is slated to become the single largest industrial contributor to climate change in North America.

By shutting down the Oil Sands and its connecting "flows of destruction" pipelines, people will have taken part in one of the most effective individual campaigns that will turn the tide of climate change.

Numerous climate justice groups around the world are dedicated to this cause. Get involved.

3. Plan Your Own Direct Action and Stand Up for Climate Justice

The Canadian and U.S. governments have failed, and continue to fail, when it comes to addressing the climate crisis. They continue to do nothing while millions are displaced or are dying due to the climate crisis. These governments have effectively prioritized destructive projects, like the Oil Sands, over the fate of humanity. The time for action is now. We can no longer allow our government to wait, stall, or block progress. Instead, we must pressure them to act and implement a just, comprehensive, and binding deal that is founded on scientific evidence, and is led by the voices of Indigenous communities and those most directly impacted by the climate crisis.

4. Help Build a Massive International Movement for Climate Justice

The greater we are, the stronger we are. Luckily, some great umbrella organizations such as the The Durban Group for Climate Justice and Rising Tide North America (and UK and Australia as well) have been working on this issue for a while now. Check out these great groups and get involved!

5. Spread the Word!

Maybe you learned something after watching the film, from reading the book, or after playing the video game. Maybe just from browsing the site you thought about things in a new way; something clicked and you had that "eureka moment" leading you from thought to action. Well if that is the case, then there is a good chance that someone in your life might learn something! Sometimes just sharing information can lead to profound changes in how we understand the world and how we want to interact within it!

To find out more information on how you can get involved, or to read up on the growing problem of climate injustices around the world, here are some tasty links to wet your social movement's lips:

THE CARBON RUSH DOCUMENTARY

www.thecarbonrush.net

CARBON CREDITS

FERN Carbon Credit Watch (http://www.fern.org/campaign/carbon-trading)

ENVIRONMENT

The Truth about Oil Sands (http://oilsandstruth.org)
Act for Climate Justice (http://www.actforclimatejustice.org/news-and-updates)

SOCIAL

Rising Tide North America (http://www.risingtidenorthamerica.org)

Climate Justice (http://www.climatejusticecoop.org)

Climate Pledge of Resistance (http://www.beyondtalk.net)

EVEN MORE INTERESTING LINKS

- http://globaljusticeecology.org
- http://www.foodfirst.org
- http://climate-connections.org
- http://www.foe.org
- http://amazonwatch.org
- http://www.mstbrazil.org
- http://www.ftierra.org/ft
- http://www.ejolt.org
- http://wrongkindofgreen.org

Amy Miller